# THE LIFE OF JAY NORWOOD DARLING

## David L. Lendt

IOWA STATE UNIVERSITY PRESS / AMES

*1979*

*For Alice, Stephanie, Tom, and Sally*

DAVID L. LENDT is Assistant to the Vice-President for Information and Development and Assistant Professor in the Department of Journalism and Mass Communication, Iowa State University. He holds the B.S. degree in technical journalism, the M.A. degree in history, and the Ph.D. degree in professional studies (higher education) from Iowa State University. He is author of *Demise of the Democracy* and editor of the second edition of *The Publicity Process*.

© 1979 The Iowa State University Press
All rights reserved

Composed and printed by
The Iowa State University Press
Ames, Iowa 50010

First edition, 1979

**Library of Congress Cataloging in Publication Data**
Lendt, David L
    Ding: the life of Jay Norwood Darling.

    Bibliography: p.
    Includes index.
    1. Darling, Jay Norwood, 1876–1962. 2. Cartoonists—United States—Biography. I. Title.
NC1429.D237L46     741'.092'4   [B]      78–10321
**ISBN 0–8138–0010–2**

# Contents

v

This is a rich, vivid biography of an enriching, vivid American. Ding Darling was an individual who loved the American earth as few have loved it, and fortunately David Lendt has caught the essence of this unusual man.

STEWART L. UDALL

# *Preface*

MOST MEMBERS of the human family are at once so complex and so commonplace that a book-length biography of any one of them would be too detailed and dull for any reader's enjoyment.

I never met Jay Norwood Darling. Several of the thousands who knew him warned me that putting Ding between the covers of a book would be a nearly insurmountable challenge. I freely admit to them that Jay Darling is not captured within the following pages. The difficult writing decisions had nothing to do with what to include but with what to leave out. And what was omitted could, in skillful editorial hands, be fashioned into several thematic tomes. Jay Darling lived fully and well. He tasted the juices of sports and politics, fame and despair, art and power, science and education, journalism and conservation. For more than eighty years he plied his talent for wringing everything possible from life.

I did meet Jay Darling's only daughter on October 21, 1976, the one-hundredth anniversary of Jay Norwood Darling's birth. At that time I knew Darling only as a Pulitzer Prize–winning cartoonist and conservationist and as a man whose life seemed to span my interests in American journalism and history. In the months that followed I studied approximately 15,000 documents; interviewed dozens of persons; reviewed newspaper files; and corresponded with friends, relatives, acquaintances, archivists, and admirers. I "lived" with Ding.

That experience brought me to two inescapable conclusions. For one, Jay Darling and I would probably have been unable to rationally discuss domestic or international policy with each other. For another, he would have been appalled at my lack of ecological sophistication and especially at my ignorance of biological science and water management. Probing the record of Darling's life also showed me that he was a fascinating combination of seemingly inconsistent beliefs and attitudes. As far as I could tell, he never wrote an uninteresting letter. He was a man of strong and stubborn conviction; a man capable of tremendous human compassion; a man who enjoyed prodigious powers of expression, uncommon energy, and immense commonsense intellect. I learned that Ding enjoyed the boundless affection of those who knew him. They were, to a person, eager to help put his life into words. This eagerness emanated from affection for the man, not from any notion that the task could be accomplished or that they would bask in Darling's glow. I discovered that the biographer's most deadly malady could infect me—that I could become the subject's friend and apologist, my inoculation of objectivity notwithstanding.

I have tried to write an unbiased account of Darling's remarkable and eventful life. The result is a reflection of Darling's values influenced as little as possible by my own. My interpretations and syntheses may be open to criticism; but they are mine, conscientiously arrived at and, I believe, wholly defensible. My editing is also subject to responsible review, and that process may result in differences of opinion concerning emphasis on one topic or lack of it on another. Whether the result speaks favorably or unfavorably about Ding to the reader, every word was written with affection. It is my belief that we are often loved as much for our human weaknesses as for our strengths. Ding would not have agreed with that, either!

Special thanks are due to Dick and Mary Koss for their confidence in me, for their candor, and for opening their home and hearts to me in my effort to become acquainted with Mary's famous father.

Sherry Fisher, executive director of the J. N. (Ding) Darling Foundation and a renowned conservationist in his own right, has been helpful since the day I decided to follow Ding's fascinating trail.

John Henry, who has devoted a significant part of his long life to the preservation of "Ding Things," has given me insights and anecdotes that only he could share.

Robert McCown and Francis Paluka were most accommodating in making available the well-maintained and carefully inventoried J. N. Darling and A. F. Allen Papers in the Department of Special Collections at the University of Iowa Library.

Jim Leonardo, custodian of the Darling Collection at the Cowles Library and a long-time scholar of Dingnalia, was gracious in providing materials of interest and value.

Every person interviewed was interested and helpful. Most of the sources requested they be given no recognition. In deference to their wishes, their names are not listed here. They do appear, however, in the Bibliography and some are mentioned in the text.

I am deeply indebted to Carl Hamilton, who is vice-president for information and development at Iowa State University and my boss. This book is the third such effort he has not only endured but aided and abetted, to the detriment of his already incredible work load.

I owe much to L. Glenn Smith, Professor of Education at Iowa State University and chairman of my graduate committee, and to the other members of that committee—Robert L. Crom, Larry H. Ebbers, Arthur M. Gowan, and James W. Schwartz—who allowed me the latitude to pursue this unusual education research project.

Kari Ellett typed the manuscript and scores of revisions with no outward sign of dismay. Nancy Bohlen, a fastidious editor, made substantial improvements throughout the book. Anita Albert designed the dust jacket and cover. I am grateful to each for her skillful contributions and patience.

Dozens of others deserve commendation and thanks. So many passed along clues, tips, names, and suggestions that listing them would run a far greater risk of omission than I am prepared to take. One informant went so far as to provide a tape-recorded interview her mother had conducted with a Ding colleague.

My greatest debt, without danger of overstatement, is to my wife Alice and our children, Stephanie, Tom, and Sally. For a decade our offspring have known Dad was taking part-time classes the year around and that full-scale family vacations and other pleasures of togetherness were out of the question. They had resigned themselves to the prospect that their father would be eternally enrolled.

Thanks, finally, go to Ding himself. His knack for arming the written word with meaning and visceral power made reading his letters, speeches, and articles an adventure. His prolific correspondence made it possible to piece together his life with few gaps or soft spots. His style adds spice to the Ding Darling story and tells a significant part of that story without accompaniment. If this biography is dull, it is certainly not Ding's fault.

Darling was a man ahead of his time in many ways. As an ecologist he is still ahead of many of us and is still adding the names of recruits to the conservation rolls. Mine is one of them.

DAVID L. LENDT

Ding: THE LIFE OF JAY NORWOOD DARLING

# 1

## *Witness to Waste*

JAY NORWOOD DARLING was born October 21, 1876. Thirty-five years later he was so famous the New York *Globe* facetiously suggested that "the Centennial at Philadelphia was laid down in the year 1876 in order to coincide with the birth of J. N. D."[1] Darling took his middle name from his birthplace—Norwood, Michigan, on Grand Traverse Bay. It was a remote and primitive area forty miles from any rail line, shut in each winter by heavy snows and vicious cold. Jay's father, a former schoolteacher and principal who had turned suddenly to preaching the gospel, had been sent to Norwood by the Michigan Methodist Conference.

The conference was to pay the Reverend Marcellus W. Darling and his wife $500 per year and provide a house for them and their son Frank Woolson Darling. In fact, the family received $69 in local scrip, and the remainder of the monetary obligation was satisfied with provisions "gladly and freely" shared by members of the congregation.

Marc Darling had enjoyed some success as a student, a professor, and an educational administrator before "politics" saw him fired from a responsible position. A native of Cattaraugus County, New York, Jay's father was born in 1844 in a one-story, unpainted frame house. He was the eighth of eleven children. One of his brothers died in 1850, another in 1851. Young Marc nearly died of scarlet fever.

Marc Darling's most memorable childhood experience, one with mystic and religious overtones, occurred about the time his two brothers died. He described the experience more than a half-century later:

My mother was sitting by the window, meditating, as was her habit, in silence, when somehow by question and answer between us, there came into my mind the sense of God's presence in my soul. A sense so real and so living that it has never left me, now after more than 56 years. I carried it with me throughout all

3

these years, in doubt at times of many other things, in distress, in battle, in college, in all my experiences.

Darling described his mother as pious, even though she rarely spoke of religion. Diantha Groves Darling was quiet, meditative, and reserved but she wielded great influence over her childen. Timothy Darling, Marc's father, was as hot-tempered as he was tenderly affectionate. The large family had difficulty making ends meet on their small farm, and Tim Darling's mercurial personality made poverty no more bearable. The Darlings wore home-sheared, homespun, and homemade clothing. Although Marc could not remember going hungry, he knew "we often came very near to it."

Marc was a dreamer in a time and place where hard work and long hours, in the Puritan tradition, brought society's rewards. He was branded as lazy because he enjoyed reading and contemplation. He read and reread the Bible and a coverless copy of a history of Greece. He traced his later appreciation for the Greek language and literature to the tattered book with which he spent so many boyhood hours. Marc Darling admitted he was a "dull boy" and that he failed to "wake up" until he was about fifteen years old.

As a young man he hired out in a succession of jobs, doing chores during the winter and working in a cider mill at least part of one summer. In September 1861—just a few months after the outbreak of the Civil War—he attended Randolph Academy with three friends. The four lived and cooked for themselves in a room over a store. Marc Darling had been bitten by the desire to learn.

In September 1862 he enlisted with his brother DeLoss for three years, or the duration of the war, in Company K of the 154th New York Volunteers. Marc fought at Chancellorsville, was captured and escaped, and retreated with the Union Army. DeLoss Darling was captured at Gettysburg July 1, 1863, while Marc was hospitalized in Philadelphia with typhoid fever. DeLoss died at Annapolis January 16, 1865.

During the winter of 1863–64 Marc studied the "School of Soldier" textbook, passed the Army officer examination, and was recommended for a major's commission in a Negro regiment, which he declined. He later joined the forced march to the relief of Knoxville and took part in Sherman's march to the sea. At Pumpkinvine Creek he became a company commander because all the officers were sick or absent, and he conversed with General Sherman under fire on the skirmish line.

Marcellus Darling harbored little respect for politicians. Following Lee's surrender at Appomattox, Darling and his Union comrades camped at Raleigh, celebrated with returning Confederate soldiers, and recounted with them the campaigns of four bloody years. "They were fully as happy as we were over the results, for the common people, soldiers, were never anxious for war," he observed, "except as the politicians and agitators roused them with appeals false and full of lies."

In August 1865 Marc Darling went to the Michigan farm home of his brother John at Albion. He worked on John's farm and that fall began the preparatory course at Albion College. He lived at the farm, which was about three miles from the school. In 1870, five years after he began his studies there, Marc graduated from Albion. That fall at the age of 26, he became superintendent of city schools at Grand Haven, Michigan, at an annual salary of $1,200. The following summer he was elected professor of Greek at Albion College and returned to his alma mater at about half his superintendent's pay—$700 per year.

Christmas Day, 1871, Marcellus Darling was married to Clara R. Woolson at Mount Pleasant, Iowa. Clara, also a graduate of Albion College, and Marc had met at the school. The couple returned to Albion, where their first son, Frank, was born ten months later. In 1874, after three years of teaching at Albion, Marc Darling took his family to Ann Arbor, where he earned the A.M. degree in philosophy in 1875. He then went to Forestville in his native New York as the principal of Regents Academy, again at an annual salary of $1,200.

Politics again entered Darling's life and brought it to an abrupt turning point. After school had closed for the academic year and everything was seemingly settled, he received a note from the secretary of the school board informing him that a new man had been elected to take his place. "Politics in the School Board did it," Darling exclaimed, and he decided to leave teaching. He later wrote, "It was then I resolved, under a sense of duty, to preach the gospel." As Jay Darling grew in his mother's womb, Marcellus Darling was groping his way through a tunnel of religious doubt. He emerged "with a tried faith in the simple, essential and practical truths of the gospel."[2]

Darling's religion was pragmatic, and his philosophy included respect for nineteenth-century American self-reliance. In 1905 he wrote:

Since the time of [Sir Francis] Bacon we have learned to apply the inductive method with very fruitful results to nearly everything else, except the Bible. We can no longer delay. We must have a doctrine of inspiration which corresponds to the facts. . . . We praise charity, and invoke heaven for a larger supply. But it is a fact to which we are beginning to wake, that this very charity without eyes is the cause of the social parasite.[3]

Shortly after the new minister and his wife were settled in Norwood, Jay Darling was born. Despite the rude conditions, that year was one of the happiest of Marc Darling's life.[4] Jay's one distinct memory of Norwood led him to judge that the community, even in its period of greatest prosperity and activity, was never an Eldorado. His mind's eye recalled the inside of the Darlings' woodshed hung with spareribs, which constituted the chief contribution to his father's salary from members of the congregation. Marc Darling received provisions in abundance, although he often had to trade eggs for flour at the grocery store.[5]

The Reverend Marcellus Darling and his family were transferred from Norwood in the fall of 1877, when Jay was only about a year old. The Darlings were sent to Cambria in the southern part of Michigan, where they remained for two years at an annual salary of $800. In May 1879 Darling accepted a call to become pastor of the First Congregational Church in Elkhart, Indiana. He went to Elkhart at a salary of $800, but his compensation had grown to $1,200 when in 1886 he moved on to the First Congregational Church of Sioux City, Iowa.[6]

While in Elkhart the Darlings became acquainted with a man who had a hand in shaping the life of Jay Darling and who launched the versatile business career of Jay's brother Frank. Lamarcus Thompson sent an illustrated "Pat and Mike" card to Marc Darling from England, where he had gone to build Europe's first roller coaster.[7] Jay was intrigued by the drawing on the card and tried to imitate it. He soon made a habit of carrying a pad of paper and a pencil and sketching several objects each day, even though the Puritan ethic was still the credo of the family. Drawing was considered a waste of time in the Darling home. "To my father and mother artists who drew pictures were classed with wicked playing cards, dancing and rum," Darling reported.[8]

Most of Jay's recollections of his youth, however, were drawn from his days in Sioux City, when that community was a robust and roughshod Missouri River town and a gateway to the unspoiled prairies of Nebraska and South Dakota. An energetic youngster of ten when he arrived in Sioux City, Jay grew to manhood there, left for college from there, and returned there to begin his newspaper career.

His youthful experiences in communities on the edge of untrammeled wilderness contributed to the creation of a zealous conservationist. Even though he left Michigan as a very young child, he could recall Indians coming to the back door with duck eggs for sale and had been told of their throwing several wild ducks into the bargain because the fowl were so plentiful.[9]

If ducks were plentiful, money was not. Darling was scolded much of his young life for making spots on the family tablecloth. "When I was a youngster and Mother did our laundry," he wrote, "I had to cover every spot I made on the tablecloth with a coin, and coins were not easily come by in those days, but even that didn't break me of the very bad habit." Even so, the family fared well enough at a time when five cents worth of liver from the local market, with baked potatoes and Hubbard squash, was enough for a Sunday dinner for a Methodist minister's family of four.[10]

Jay Darling thrived in the spacious fields of high grass through which he and his brother Frank roamed, whether afoot or astride their spirited Indian ponies. He spent many summer nights on the prairie or along the banks of the Missouri River or the Big Sioux River, listening to the voice of the puma—or panther, as it was called in northwestern Iowa. He intended in those days not to let any marsh, lake, or pothole escape his attention. He came to know eastern South Dakota as well as most youngsters know their backyards; and his communion with South Dakota's pristine bounty—its head-high grass and abundant underground water—provided the ingredients of the "pleasantest recollections" of his long and eventful life.[11]

One of Jay's Sioux City neighbors was Max McGraw, who later headed the huge McGraw-Edison Electric Company. Jay, the preacher's son, was obliged to take an active part in the church choir and to play the organ. McGraw was active in a supporting role. "I little thought," he later wrote, "when I was a boy pumping the organ in the church of Ding's father that either of us would see the changes in this world of ours which Ding so meticulously" recorded in his cartoons.[12] The lives of the two boys converged on the banks of the Missouri, and the lives of the two men later converged in common conservation interests.

During some summers Jay hired out to his Uncle John in Albion, Michigan, where he mowed marsh hay with a scythe and cradled the wheat in the corners of the south forty that the McCormick reaper could not reach. The work was strenuous, but the yield was bountiful. The small but rich farm was a paradise to the teenaged Darling. A clear stream ran across one corner, and a small stand of native timber stood on virgin soil. The water was full of fish, migratory waterfowl nested beside the stream, and the timber was alive with songbirds. He fished in the same creek and hunted on the same millpond and timbered shores near Albion as did Lynn Bogue Hunt, who became famous as a wildlife artist.[13]

Jay Darling's first lesson in conservation came literally at the hand of his Uncle John. The youth had shot a wood duck in midnesting season, and John had blistered his nephew's rear. In the proper season, however, ducks were so abundant that Jay recalled his uncle saying, "Now, I will do your milking for you; you go down and get a mess of ducks for dinner."[14]

In other summers, after he was old enough to be "turned loose," he rode out of Sioux City, across the Big Sioux, and into South Dakota to herd cattle for anyone who would hire him. "Those were the days when the Golden Plover came in great flocks and moved across South Dakota and[,] from early spring until the Prairie Chicken sought cover in the fall along the thickets bordering the creeks and marshes[,] my mind was filled with pictures which have never been erased," Darling wrote. "It was the disappearance of all that wonderful endowment of wild life which stirred the first instincts I can remember of conservation."[15]

When the Darlings first moved to Iowa, flat-bottom stern-wheelers plied the waters of the Missouri River between Sioux City and Kansas City, even though "they spent most of their time stuck on sand bars." About 1885 the last such riverboat ran onto a snag, ruptured its hull, and sank. As a boy, Jay saw the stern-wheelers loaded with bales of raw buffalo hides for the Saint Louis market.[16]

Marcellus Darling was not a wealthy man, but he believed deeply in the advantages of education. In 1895, when Jay was nineteen years old, the Darling family toured Europe. The excursion was made on $400 per person, for which every Darling had "worked, saved, read evenings and planned for years in advance." In 1888 Jay Darling earned a 4 x 5 wood box camera by selling ten subscriptions to *Youth's Companion* and scratching up $5 in cash, and he had become proficient as a photographer. His was the first amateur camera in Sioux City, and with it he took pictures of his inseparable companion, Jay Dundass; his family; friends; and the neighborhood around 1423 Nebraska Street. The same fourteen-inch-long camera accompanied the Darlings on their European tour, "greatly to the discomfort of all—particularly the endless customs officers." The elder Darling also set aside the resources and the time for summer vacations. Jay Darling recalled the Storm lake, Iowa, resort of the early 1890s and Mae Beggs. The Reverend Marcellus Darling and Doctor Beggs brought their children to Storm Lake and, whether by accident or design, kindled Jay's first romance. "Mae Beggs," he later wrote, "blond, beautiful and dumb—who furnished all the dreams and aspirations of my callow days."[17]

Darling also retained more sobering recollections. He remembered the day Sioux Citians gathered on the riverfront to watch the city of Covington, Nebraska, on the opposite bank of the Missouri, crumble into the river. The Missouri river had undercut the community's buildings and destroyed them "while the people skedaddled to the hills."[18]

Perhaps even more clearly, Darling remembered the shattering experience of returning to his uncle's farm as the family representative to John's funeral. "It was the first time I had seen my youthful paradise since I was about fifteen years old and it seemed as if the farm had died with Uncle John," Darling wrote. He strolled down the lane to the pasture and the woods. The topsoil of the grain field had been stripped away. The timber had been cut. The river was reduced to a muddy trickle unfit for game fish. The pasture, bare of grass, was scarred and slashed by erosion and no longer of any use. A solitary crow rose from the barnyard and slowly flapped out of sight—the only sign of wildlife left on the place. The well had gone dry. The orchard was a tangle of dead limbs and tree stumps. There was nothing left worth cultivating where Darling had once plowed eight inches of black loam and reaped sixty bushels of wheat per acre. "This was my first conscious realization of what could happen to land, what could happen to clear running streams, what could happen to bird life and human life when the common laws of Mother Nature were disregarded," he later wrote.[19]

Sioux City and its surroundings were still new, and humankind had not sullied them. The rivers were deep, and game of all kinds was abundant. "Life," Darling recalled, "was rich and easy. Everyone hunted and everyone fished and everyone helped himself liberally to the riches which Nature provided." But by the time Darling himself was a father, waves of settlers had plowed up the prairies, cut the timber bordering the creeks and rivers, harvested the game, and turned their cattle loose to strip the protective grasses from the land. The unconcerned pioneers had spoiled their eastern nests and pushed westward to abuse newer and greener pastures.

Darling brooded at the prospect of seeing the saga of his Uncle John's farm repeated. He read of similar concerns expressed much earlier by Benjamin Franklin, Thomas Jefferson, Audubon, and Samuel Adams. Theodore Roosevelt and Gifford Pinchot later fired his determination with their courageous endorsement of conservation principles and their condemnation of those who violated the tenets of good stewardship. The minister's son was so inspired that he "spoke up in prayer meeting" for restraint, even though most of the parishioners were convinced "that Uncle Sam was rich enough to give us all a farm, and let us get ours while the getting is good."[20]

"If I could put together all the virgin landscapes which I knew in my youth and show what has happened to them in one generation it would be the best object lesson in conservation that could be printed,'' Darling claimed. He later wrote, ''All it takes to be a conservationist is to have been awake and a witness to what has happened to all our continental forests, soils, waters, minerals and wildlife in the last fifty or seventy-five years and he'll be a conservationist from fright! That's me.''[21]

Jay Darling, a child of the centennial, had seen that just one generation of indifference, neglect, and needless abuse of natural riches could shortchange every generation to follow. He saw the results of despoliation. He was a witness to waste.

# 2

## Lamp of Learning

IT HAD BEEN understood that when Jay Darling reached college age he would go "back east" to Albion College, where his mother and father had met and been educated and where his father had taught Greek and Latin. As the time for his departure neared, however, Jay changed his mind.[1]

He began his checkered college career at Yankton College in South Dakota in 1894, just before his eighteenth birthday. There he vented some of his youthful energies in sports and some in other extracurricular activities. His leadership was evident as he organized, captained, and played on the school's first football team. After a year, however, he was sent home from the small Congregational school for his less constructive activities. He and some friends had "borrowed" the college president's horse and buggy for the evening. The next day the president noticed the horse's stiff-legged gait at about the same time he discovered Darling's cuff link on the floor of the buggy. It was the end of Darling's career at Yankton.[2]

The following year the irrepressible Jay began again. With several Sioux City friends he enrolled at Beloit College in Wisconsin, where he hoped to prepare himself for medical school. He wanted to become a doctor, "like his red-faced Uncle Frank, of Jamestown, New York."[3]

He was active at Beloit too. Ordinary pastimes, however, could not slake his thirst for the extraordinary, the unique experience. With a classmate he bicycled from Beloit to the Chicago University football field. On the way the two stopped at a creamery for buttermilk and napped with bedbugs at an old railroad flophouse. As they neared the end of the 100-mile journey, Darling's legs rebelled against the unusual task put upon them, and it took nearly a week for him to recover from the exertion. Darling claimed he never got on a bicycle again until the gas shortage of World War II.[4]

1 1

At Beloit, Jay paid much of his own way with his considerable musical talent. He was an accomplished mandolin player and bass vocalist. "I could sing in any religion you wanted," Darling reported, "and I made the rounds of all the funerals every week." He sang and played for parties and for graduations, and he sang and played for the fun of it. His roommate at the Beta Sigma Psi house at Beloit was Paul C. Howe, a fellow musician who became Darling's lifelong friend. Darling, who was not much of a joiner, was proud of his fraternity affiliation. His friendship with Paul Howe, Darling wrote to Howe's son, "runs through my life like the theme of a symphony." Howe played guitar and banjo and sang with gusto. The twosome would occasionally spend an entire evening crooning the current hits and drinking beer, to the delight of their appreciative friends.[5]

By the time he was a junior at Beloit, Darling had become leader of the glee club, manager of the track team, managing editor of the college paper, a member of the mandolin club, a regular member of the Methodist Episcopal church choir, and the steady escort of Miss Bit Sumner. He was also regularly late to the president's 8:00 A.M. class in Evidences of Christianity, he was flunking all his courses except biology,[6] and he was in trouble for his behavior in Greek literature. The teacher, Theodore Lyman Wright, sent Darling a handwritten note:

I can scarcely doubt that you are by this time sorry, as I am, that the good order of a Beloit College class was so violently interfered with by the manner in which you left the Greek Literature recitation last Friday. Of course you will not expect to be again admitted to any exercises in my room until you have made the past right with me and with the class. I am extremely sorry if you find my manner in the classroom irritating, and I can assure you that 'tis not intentionally so.[7]

The minutes of Beloit's faculty meetings also reveal that on January 21, 1898, Darling and a Mr. Jeffris "were suspended until Monday night, or longer if their pieces should not be in by that time," as punishment for delays in their rhetorical assignments.

As a junior, Jay was also art director for the Beloit yearbook, the *Codex*. His flair for sketching contributed to his academic downfall and to his first use of the signature by which he became nationally famous. Near the turn of the century Beloit College was not a center of liberality or of familiarity between students and faculty members. The Congregational school expected exemplary behavior from the sons of ministers placed in its tutelage. Darling's performance was less exemplary than exceptional.

Darling repeatedly claimed he did not know why he was chosen art

editor unless it was a joke. Once in charge, however, he chose to illustrate the *Codex* with sketches of the most staid and stern of Beloit's
faculty, including its president. Although the sketches were rudimen-
tary, Darling's skill with the pen was sufficient to make the subjects
easily recognizable—whether as the Devil shoveling students into a fiery
furnace or as a scrawny male ballet dancer attired in a tutu. The drawings, in fact, were so well done that the members of the small, reserved
faculty considered the publication shocking. They did not appreciate
the sketch of a revered professor of Latin singing, "There'll Be a Hot
Time in the Old Town Tonight." They also failed to see the humor in
the sight of the president of the college, E. D. Eaton, dancing the
Highland fling in appropriate attire. They did not like seeing the principal of the academy, a devoted soldier of temperance, singing, "Give
Us a Drink, Bartender." The caricatures of Prexy and Tommy and Billy
and Dutchy—all serious Congregational college professors—were dis-
concerting enough. The sketch of faculty lined up as chorus girls was too
much for their dignity to bear.[8]

Darling signed his creations D'ing, a contraction of his last name.
"The apostrophe stood for the 'arl' which were left out in order to make
a funnier looking signature and in addition to conceal my identity,"
Darling explained. The name had been used by his father and by his
brother Frank before him. And Ding had become his nickname at
Beloit.[9]

A vote was taken by the indignant faculty and, predictably, Darling was suspended from Beloit for one year. Circulated and given
credence by the popular press, Ding folklore for years held that Darling
was expelled from Beloit strictly for his performance as art editor of the
yearbook. Darling himself set the record straight in a speech on the
Beloit campus in the 1950s when he said, "The truth is I was not fired
because of that picture, but because I flunked damn near every study I
took that year. I was a no-good student and that's why I graduated in
1900 and not 1899." The faculty minutes confirm Darling's account;
June 17, 1898—"It was voted that Mr. Darling be suspended for a year
on account of irregularities in attendance and poor scholarship."
Nothing was said of the president or members of the faculty in tutus,
said Beloit's archivist, Robert H. Irrmann. "Obviously, they hung him
with a different rope!"[10]

Suspension from the church school brought with it an interlude of
humiliation for the son of a respected Sioux City clergyman. Darling
was not ready to face his parents or his many friends in his hometown.
That summer he barnstormed the circuit of Chautauquas with a male
quartette. During his year of suspension he also visited Florida for the

first time and was moved by the state's unspoiled tropical swampland, its coastal islands, and its teeming wildlife. That fall he slept in an unheated attic, got up before dawn to milk about a dozen cows before breakfast, and picked eighty to one-hundred bushels of corn every day on a farm about two miles from Castana, Iowa. Near the end of his enforced vacation from Beloit, Darling returned to Sioux City for his first taste of newspapering. For a short time he was a reporter for the Sioux City *Tribune,* the local competitor of the Sioux City *Journal* where he later began his career as a cartoonist.[11]

Darling was capable of making his grades at Beloit, but he chose his own scintillating course, declared his independence, and accepted the consequences. He would not be constrained by the conventions of parochial higher education nor by prescribed programs of study. He apparently applied himself to his studies in the life sciences, where his interest and his fascination were attracted, and gave short shrift to most other academic requirements. Yet his lifelong instincts were scholarly and scientific. He observed, studied, and researched the world around him. His quick mind was eager to receive the stimuli surrounding him and to process, store, and assimilate them. He continually exercised his mind as he delved into an array of interests and avocations. He abhorred superficiality and insisted on probing beneath the familiar exterior of a topic. He believed that any reasonably intelligent person could become an expert in any area with concentrated effort over a short time.

Darling eventually returned to Beloit for his senior year of study and graduated in 1900. Even though his academic performance was not memorable for its excellence, his experience at Beloit influenced his life significantly. Darling was especially thankful to one teacher for having stirred an interest in the broad field of biology. Because of that one professor's influence, Darling thereafter saw the world as a biological system and viewed the world's events against that ecological background. He was interested in medicine and thus in human life. But he was led to understand that human life cannot exist divorced from the plant and animal life with which it shares the earth's soil, water, and air. Nearly a half-century after his graduation, Darling wrote, "I have been a long time out of school but I still remember that I had one teacher who applied his course in biology to the functional relationship of biology to man's existence."[12]

Darling still aimed to become a man of medicine, but he had glimpsed his future patients as members of one species in a broad, interrelated collage of life. His respect for all living things had been heightened; his desire to preserve life had become his obligation.

# 3

## A Critical Shift

ABOUT THE TIME Jay Darling returned to Sioux City as a graduate of Beloit College, his father and mother left town. Marcellus Darling attributed the move to asthmatic trouble and wrote that the "happiest possible relations existed when we left, March 1, 1900." The Darlings traveled to Glencoe, Illinois, where they took charge of the Congregational church the day following their arrival.[1]

Jay kept his eye trained on a career in medicine. He wanted to be "a doctor with a cheery, red face and a rotund stomach and flowing side whiskers and a tall silk hat," but he knew it was asking too much to request a course at a medical college.[2] He needed to work and put aside some money.

He joined the Sioux City *Journal* as a cub reporter under the direction of thirty-three-year-old Arthur Francis Allen, the paper's managing editor. John W. Carey, just a few months younger than the twenty-three-year-old Darling, was also a reporter for the *Journal*. Darling's Beta brother Paul Howe was a *Journal* employee, as was Fred W. Beckman, who later headed the journalism program at Iowa State College. George D. Perkins, the founder of the paper and its hard-bitten editor, was a man of principle in a town where there was little popular demand for high standards of behavior.[3]

Sioux City in 1900 was a wide-open town featuring beer, liquor, and wine sales on Sunday and a proliferation of saloons, public gambling places, and houses of prostitution. The community rested on the fringe of the frontier and many of its residents harbored a crude and violent frontier spirit. Only Covington, Nebraska, was rougher. It was "the toughest spot in the northwest in those days," according to Darling, and was connected to Sioux City by a pontoon bridge without railings. One of Darling's earliest assignments as a reporter was to rise early in the morning and join the coroner and the deputy sheriff in a lumber wagon to patrol the shores of the river below the bridge searching for bodies. Unwary or unstable adventurers making the crossing were frequently knocked unconscious, robbed of their valuables, and dumped from the bridge. Several beachcombers lived in shacks along the banks of the Missouri and on a sandbar near the middle of the river. They made their living by fishing bodies from the water for one dollar each. Darling recalled, "We would pick up a stiff about once out of three mornings and it was always good for a story—especially when the corpse could be identified. . . ."[4]

Even in such an atmosphere editor Perkins kept a firm hand on the throttle, and he expected his employees to hold to his standards. The rules Perkins laid down for his reporters were remarkable for their time. *Journal* employees were forbidden to accept free access to or free membership in just about anything but a local church. No *Journal* employee could accept a complimentary membership in any of the local boat clubs. Perkins preached constantly that free tickets to anything, even the railroad, warped the recipient's mind until he could not write a fair story on the subject. Darling counted himself fortunate to have learned the business under George Perkins before going out into the newspaper world.[5]

Darling's life in Sioux City was one of contrasts. During the week he wrote of bodies violently disposed of in the neighboring river, and on Sundays he was choir director and bass soloist for Saint Thomas Episcopal Church. He held the dual position for about two years, probably out of a sense of gratitude for what the Episcopal choir had done for him. Even though his father had become a Congregational minister, Jay joined the Episcopal choir at the age of twelve and considered it the first memory of any important assignment in his life. Darling recalled that the parishioners "said it was a good choir, but I doubt it. Anyway, it was a good music lesson for me." On one occasion he ran a summer encampment for choir boys at Brown's Lake and was nearly fired from the *Journal* for his week-long absence. When he finally announced his decision to resign as choir director, Darling received the following penned note:

We, your friends and admirers in St. Thomas' Church parish, beg you to reconsider your determination to sever your connection with the choir. Your success in this work has given universal satisfaction and we sincerely hope that you will signify your willingness to remain with us as choir director.

The note was signed by thirty-four persons but was to no avail.[6]

On at least one occasion Jay's interest in the church and in the

*Journal* coincided. The resourceful young reporter hid in the organ loft of the old Congregational church and listened in on "a sensational church meeting that was supposed to be strictly private. . . ."[7]

While he listened and wrote, Darling continued to sketch. He studied the features of local citizens and reduced them to a few lines on random sheets of notepaper. The drawers of his desk at the *Journal* office became cluttered with bits and beginnings of simple portraits of many of the most familiar residents of Sioux City. Even so, drawing remained no more than a sideline with him until he was chased into his meteoric cartooning career by a crotchety attorney who objected to having his photograph taken.

Darling's interest in photography, dating from his introduction to the *Youth's Companion* box camera, had probably helped him get his job at the *Journal* at a time when newspaper photography was in its infancy. He had been covering a district court trial and had written a lively account of the attorney's animated verbal and physical attack on the counsel for the opposition. Darling showed his lengthy story to the city editor and was instructed to get a photograph to go with it. Darling dutifully lugged the primitive newspaper camera down the street, stalking the temperamental attorney. When he found the lawyer, who was still in the courtroom, Darling tried to snap a photograph secretly. The barrister spotted the newsman, let out a whoop, jumped over a chair, and took after the young reporter, swinging his cane in the general area of Darling's head.

"I didn't emit a whoop," Darling recounted, "but I leaped over another chair and started for the door. The lawyer was close to me as I sailed down the steps in one leap, and the air from his cane breezed across my neck. On the street he was no match for me because my legs were young and his were old, so I soon outdistanced him." Darling escaped without injury and without the photograph he had been ordered to produce. He returned to the *Journal* office and began rummaging in his desk among the scraps and sheets of notepaper. There he found a sketch of the agitated attorney. The editor liked what he saw and ran Darling's artwork with the story.

When the drawing appeared the next day, it enjoyed such a favorable reception from *Journal* readers that Darling was soon assigned to draw a succession of sketches of Sioux City characters. For the first series, titled "Local Snapshots," Darling drew caricatures of well-known citizens in familiar clothing and surroundings. The series was a popular success. Later, in cooperation with John W. Carey, Darling illustrated a series of feature articles titled, "Interviews that Never Happened." Several of Ding's series were well-read vehicles for poking generally harmless fun at some of the pillars of the community. They may also have cost Darling his position at the *Journal*.[8]

Darling sketched his first conservation cartoon during Theodore Roosevelt's term as president, which began in September 1901 when President William McKinley died. The cartoon backed up Roosevelt's campaign for the establishment of a forestry service, which had been given birth and wide national emphasis by the man who became the first Chief of Forestry, Gifford Pinchot. Years later Darling recalled that Pinchot carried on a continuing battle for forest preservation against his superior, Secretary of the Interior Ballinger who was inherited by Roosevelt from the McKinley administration. (This was one of the rare instances in which Darling recounted the past incorrectly. Richard Ballinger was secretary of the interior under President William Howard Taft, who succeeded Theodore Roosevelt as president.) Roosevelt credited Pinchot with being the "father of conservation," Darling recounted, "and I think Teddy Roosevelt learned most of his forestry lessons from Pinchot, who later became Governor of Pennsylvania, and I am sorry to say not a very good one."[9]

Even though Darling's career as a reporter-cartoonist was well under way, he was not distracted from his many avocations, foremost of which were hunting and fishing on the thousands of acres of unspoiled prairies and wetlands west of Sioux City. Years later even he was amazed at the pace he had maintained in his twenties. "I don't know how I got along with so little sleep in those days," he wrote.

I was both cartoonist and reporter on the old *Journal*. Those were the days of open spring shooting and long before there was any such thing as an eight-hour day. It was a regular routine to put the paper to bed about midnight, pile into the buckboard from the neighboring livery stable and head for the flats along the Missouri River and get back in time to cover the police courts for the afternoon paper, draw a cartoon and cover local assignments until the next morning's *Journal* was put to bed.[10]

Darling camped for a week at Leech Lake with Russell Marks of Sioux City; and he traveled the marshes, lakes, and potholes of eastern South Dakota, where he heard now and then about a lad named Errington who could shoot "like nobody's business." The young man was Paul Errington, who became part of a daring Darling experiment in education thirty years later.[11]

When the venerable George Perkins, the *Journal*'s founder, made a run for the governor's chair, Darling campaigned for him through *Journal* cartoons. The *Journal*'s good reputation and wide circulation insured Darling's fame throughout the state. The cartoonist began to hear tempting offers from other papers in Iowa.[12]

Darling was active on other fronts also. He was particularly interested in a young Sioux City woman named Genevieve Pendleton. His Beloit classmate Paul Howe was courting Mary Weare Nason, a young woman who had moved to Sioux City about 1904 to live with her grandparents. Darling and Howe, ever the minstrels, hired an open hack one moonlit night and went about Sioux City serenading their girlfriends, embarrassing both. Miss Pendleton, known as Penny, was one of three orphaned children of Isaak Pendleton and Margaret McDonald Pendleton. Penny's uncle and aunt, John M. and Nettie McDonald of Sioux City, raised the young woman. Uncle John McDonald was a banker and a sheriff of Woodbury County "in the tough old days."[13]

As the relationship between Jay and Penny grew, so did Darling's reputation. Early in 1906 W. H. Powell of the Ottumwa *Courier* wrote Darling a note suggesting that an enclosed clipping from the Washington *Press* "should give you an ambition as high as a skyscraper." The clipping, taken from the editorial page of the *Press*, said: "Ding, the cartoonist on the Sioux City *Journal*, does some work as good and as witty as McCutcheon's. His group of pictures of Gov. Cummins reading his message is hot stuff. He has talent that will land him on a high place."[14] The message was prophetic. The renowned John T. McCutcheon, cartoonist for the Chicago *Tribune*, later became one of Darling's fans.

Darling continued to receive feelers from other Iowa papers, including the Des Moines *Capital*, major competitor of the Des Moines *Register and Leader*. The *Capital*'s managing editor, W. T. Buchanan, in September 1906 wrote Darling this terse inquiry: "Have you any thoughts of leaving Sioux City for a broader field?"[15]

Jay Darling was a sought-after reporter-cartoonist, although his drawing far surpassed his spelling. Penny Pendleton was amused at the misspellings in Darling's notes and letters to her and assumed they were intentional. She soon learned otherwise. He used words such as "poluter" and "omnoverous" and "Herculese" and "proffs" and "presincts" and "socety." Darling was aware of his weakness but blunted attempts to improve it. He often suggested that anyone who could not spell a word at least two different ways did not show much imagination.[16]

Darling stayed with the *Journal* for six years and had achieved "the monumental salary of $27.50 a week, on which meagre stipend I thought it was time to get married and chose one of the old classmates in public school." Jay Darling and Genevieve Pendleton were married

October 31, 1906, in the spacious home of Uncle John and Aunt Nettie McDonald. The ceremony, performed by Marcellus Darling, followed the wedding of Paul Howe and Mary Nason by about five months. While Jay and Penny honeymooned in the West Indies, a telegram arrived from the Des Moines *Register and Leader* offering the cartoonist a position there.[17] It was a good thing for Darling that it did.

A row within the school board in Sioux City had provided grist for an early installment of "Local Snapshots." The subject of the feature was the president of the board, a Dr. Dunlavy. The cartoon chided Dunlavy for his patronizing attitude toward children in the Sioux City schools. Darling's cartoon pictured the good doctor supporting an umbrella beneath which the schoolboys and schoolgirls clustered about his knees. Later, the picture apparently offended Dunlavy and Perkins, and the publisher reportedly fired Darling. Darling and Howe thought it was a great joke, especially considering Jay's offers from other newspapers across the state.[18]

The Darlings interrupted their honeymoon to return to Sioux City and make arrangements for their move to Des Moines. Although Darling was something of a Sioux City celebrity, the couple's return was not big news in the *Journal*. Someone, however, "got up" a fake reproduction of the front page of the *Journal* on which appeared a banner headline and a major story announcing the presence of the newlyweds in Sioux City.[19] The *Journal* staff, in all likelihood, did not know that Darling had made a contract with the *Register and Leader* and that he was to begin his work in Des Moines almost immediately.

Attraction to the lofty profession of medicine still tugged at Darling's innards, but Ding knew as he set out for his new position in Des Moines that he would never be Jay Norwood Darling, M.D.

# 4

## A Rising Star

JAY DARLING had signed a contract whose provisions, in his estimation, were generous. He had agreed to work for Gardner Cowles, publisher of the Des Moines *Register and Leader,* for $50 a week—nearly double the top salary he had received in 1906 at the *Journal* in Sioux City. The contract spelled out Darling's several duties and recognized for the first time that Ding was a cartoonist. According to the written agreement signed by the cartoonist and the publisher, Darling's chief responsibility was to provide an editorial cartoon each day. His other duties were subordinate to his artistic ones.

Gardner Cowles and Jay Darling, whether it was apparent to them as early as 1906 or not, had much in common. Both were sons of nineteenth-century Protestant ministers. As boys both had been required to do summer farm work. Both were raised on the edge of poverty. Both were nurtured on the Protestant work ethic and the credo of self-reliance. Both were energetic and diligent in pursuit of their goals. Both were unyielding perfectionists who expected superior performances from themselves and from others. Although staunchly independent, both had been college fraternity men. Both had been married in ceremonies performed by their fathers. They were fiscal and political conservatives. They were movers and shakers and conspicuous overachievers.[1]

Their paths crossed at a critical time in their lives. Cowles, at forty-five, had only recently given up the comfortable life of a country banker in Algona, Iowa, to throw himself into the salvation of the crippled and unsteady *Register and Leader.* Working with Harvey Ingham, who had been editor of the *Upper Des Moines* in Algona, Cowles was trying to get the *Register and Leader* on a competitive footing in spite of serious morale problems, high employee turnover, low salaries, and financial difficulties. Just two years earlier he had moved his wife and six children to Des Moines as part of his commitment to the challenging business venture.[2]

Darling, at thirty, had also made major commitments to a bride, to family life, and to the struggling *Register and Leader*. He and Penny had moved their belongings to Des Moines and, while Darling had no cash investment in the newspaper that employed him, his future required that the *Register and Leader* become a widely read and respected vehicle for his work.

One of Darling's first acquaintances in Des Moines was Vernon L. Clark, then a twenty-one-year-old employee in the *Register and Leader*'s advertising department. Clark met Darling "practically as soon as he swung down off the train," and they became close friends. Clark sold want ads and commercial display advertising for the newspaper and found Darling a valuable ally. Occasionally, he would ask Darling to illustrate an ad layout. Clark discovered that the layouts sold easily with the Darling touch, and his sales of space increased. The newspaper's new cartoonist provided his artwork free of charge to Clark. The advertising salesman ingratiated himself with clients who asked him to "get Mr. Darling to draw me a picture" of some kind. Mr. Darling was happy to oblige if it would make life easier for Mr. Clark.

Clark and Darling were together at the *Register and Leader* only about three years. Clark recalls handling the advertising for Younkers, then as now the premier downtown Des Moines department store. The advertising manager for Younkers had a drinking problem, and his employers wanted him to take a couple of weeks to get dried out because, as Clark emphasized, "he was a nice old cookie." Clark was hired for $7.50 a week to proofread the store's advertising copy. Two weeks ran into six months, and by the time the advertising manager returned to Younkers, Clark had long since been replaced at the *Register and Leader*. The *Register and Leader*'s archcompetitor, the Des Moines *Capital*, offered Clark a job in the hope that the spirited young adman might increase the *Capital*'s volume of Younkers advertising. Clark had a run-in with the *Capital*'s aggressive new advertising manager, who would not go home until the Younkers ad was in, and Clark found himself again on the street. He promptly went into the flour milling business and later the sawmill business, where he was very successful. Even though their newspaper days together were few, Clark and Darling remained friends until Darling's death fifty-six years after their first meeting.

Darling's first cartoon for the *Register and Leader* did not make his welcome to Des Moines a warm one. The drawing featured a pudgy monk smoking a pipe labeled "soft coal." The cartoon had no more than appeared in the December 9, 1906, issue of the paper than the criticism began. Catholics were offended at the notion that the monk was responsible for the heavily polluted air in the city. Darling had mistakenly assumed that "Des Moines" was derived from the French word for monk. Des Moines coal dealers were also angry, although that troubled the pollution-conscious Darling less than the fact that he had offended members of a religious denomination.[3]

Darling's first political cartoon was rendered for the Sioux City *Journal*, when Theodore Roosevelt was running for vice-president on the Republican ticket with William McKinley. Titled "An Incident of Bryan's Trip to Kansas City," the cartoon pictured McKinley and the hero of San Juan Hill aboard an elephant, confronting Bryan and "Mother Democracy" on a mule. It appeared June 27, 1900.[4] McKinley and Roosevelt were elected; McKinley was assassinated in September, 1901, thus elevating Roosevelt to the presidency.

By the time Darling was executing cartoons for the *Register and Leader,* Roosevelt was in the thick of political activity and Ding was making the most of it. Roosevelt not only was a political personality but he had found in Jay N. Darling a willing and able recruit to the cause of conservation. Roosevelt was a staunch conservationist and a founder of the venerable Boone and Crockett Club, an organization devoted to the preservation of game in particular. Roosevelt had spent several years as a rancher in the Badlands of western North Dakota, where he had been inspired by the unmarred beauty of the wild country and its inhabitants. "It is possible," one author asserts, "with only slight exaggeration, to date the beginning of the organized conservation movement in North America from the time of the arrival of Theodore Roosevelt in the West." During his stay in North Dakota, Roosevelt saw the impact of civilization on a wild area and the experience left a deep impression on his character.[5] Jay Darling was an unabashed Teddy Roosevelt fan and later became a friend of the Republican President Roosevelt.

At the time of the 1908 presidential election, when Darling was doing his drawing for the *Register and Leader,* his hero Roosevelt was stewing in a political dilemma of his own making. Teddy considered his nearly full term following McKinley's death as his first term and publicly announced he would not seek reelection following his second term. He later revealed privately that he was sorry he had made that announcement. Nevertheless, he designated William Howard Taft as his choice for the Republican nomination in 1908. Darling made the most of Roosevelt's earlier indecision, however, showing him at the end of the diving board at the old swimming hole, nonchalantly dipping his toes in the water while other cautious, would-be contenders lined up on the board behind him to ask, "Will You, or Will You Not, TR?"[6]

Jay Darling grew more expert with his pen, whether it was pointed at the political scene or the problems of conservation and pollution. "Ding" was becoming a household word and his contribution to the front page of every issue of the *Register and Leader* was becoming a trademark of the paper. By order of Gardner Cowles, Jay Darling was given free rein in his expression, without consultation with or consent of the editorial staff. It was the Cowles way of managing his newspaper enterprise, and Darling thrived in an atmosphere of independence. Just as the *Journal* had provided Darling an Iowa stage upon which to display his talents, the *Register and Leader* allowed him to play to a wider audience. He was soon being noticed, and he was being lured by other papers, including the Baltimore *Sun*, the Milwaukee *Journal,* the Chicago *Daily Examiner,* and the Cleveland *Leader.*[7]

Darling had also become a father since his arrival in Des Moines. His son John was born December 30, 1909.[8] Ding was restless and eager to explore opportunities elsewhere. He had met the challenge in Des Moines with the *Register and Leader.* He had attracted a devoted following in about four years. He knew he was in demand and believed he owed it to himself and his family to leave the Algonquin Apartments at Fifteenth and Pleasant streets in Des Moines and to extend his reach. He decided to join the New York *Globe.*

New York was "The Big Apple," and Darling knew if he could make it there he could make it anywhere. He would never know whether he could succeed as a national cartoonist unless he gave it a try. His audience would be greatly increased not only because the *Globe* boasted a larger circulation than the *Register and Leader* but because it was launching a new syndicated service through which Darling's cartoons could be marketed to client newspapers across the nation. The appeal of the syndicate had helped tip the scales in favor of the *Globe.* Darling was also amazed at the large salary the newspaper was willing to pay its new cartoonist. The cartoon on the front page of the *Register and Leader* of November 2, 1911, was drawn by Frank Moser. It bade farewell to Ding, "who's going to the New York *Globe.*"[9]

The Darlings were on the brink of an exciting expedition into the "big time." Jay's hopes were high. His self-confidence and optimism soared as he considered the expansive possibilities offered by his new position. Ding could not know that his tenure at the *Globe* would be one of the most difficult and disheartening periods of his ebullient life.

# 5

## *Hard Times*

THE NEW YORK *Globe* was near the heart of things in 1911. Jay Darling, as a member of the *Globe* team, worked alongside cartoonist H. T. (Webby) Webster in the art room on Dye Street and rubbed elbows with the editorial staff housed in the tower of the Singer Building. He grew fond of luncheon at White's Restaurant and breakfasts of English muffins and Greek honey served up at the Algonquin Hotel. The *Globe*'s masthead fairly sparkled with the names of luminaries in politics, literature, and journalism. Edna Ferber, William Jennings Bryan, and William Allen White regularly took special assignments for the newspaper. It was heady stuff for an upstart cartoonist from Sioux City, Iowa, by way of Des Moines.[1]

The gregarious Darling made friends easily, and in what was to be a short and disillusioning stay in New York he made contacts with famous and influential persons whose friendship he cultivated throughout his life. It was probably during his career at the *Globe* that Darling made the acquaintance of Henry Ford, who invited the affable young artist to lunch at the Ford home. The two spent several hours surveying Ford's collection of watches, and Ford finally asked his guest if he had a watch. "I immediately stuck out my wrist," Darling recalled, "and noting his disappointment I reached into my seldom-used but always available for emergency time regulation pocket" where he kept "one of those large model old-fashioned railroad man's watches—a $150 Hamilton" purchased in the blush of his newfound affluence. Ford checked the watch carefully, pronounced it accurate within about twenty seconds per year, and said he would not attempt to regulate it any better than that. The watch, Darling observed many years later, performed exactly as Ford predicted it would.[2]

Comfortably situated at Sea Cliff, Long Island, Jay and Penny awaited the birth of their second child. Mary's arrival October 20, 1912, was the zenith of the Darlings' unhappy experiment with living in the supercharged atmosphere of the nation's largest metropolis.[3] Edna Ferber, a Darling enthusiast, wrote a note of congratulation following Mary's birth: "Just when I had carefully and reverently packed you away in lavender, and tied you up in pink ribbons, in the firm and sorrowful belief that you had forgotten the likes av [*sic*] me, up you bob, twice in the same place, bursting the pink ribbons, and scattering the lavender all over the place."[4]

Although he was being handsomely paid, Darling was uneasy in his high position. He felt closed in and partitioned off. The newspaper business in the towers of New York was an impersonal profession when compared with the earthy occupation of journalism as practiced in Iowa. Darling was rankled by the *Globe*'s management and the tendency of its editors to suggest cartoon subject matter consistent with the newspaper's editorial views. He also resisted management pressure to do comic strips for the *Globe*.[5]

Darling was also having physical difficulties that threatened to destroy his career as an editorial cartoonist. As a youth he had crushed his right elbow in a fall from a horse. The broken bones had never healed perfectly, and just as the young artist's career seemed about to soar, the elbow grew increasingly painful and his valuable drawing arm grew less controllable. Even before he had decided to make the move from Des Moines to New York, Darling was hospitalized in Chicago, where physicians attempted to diagnose and correct the malady. Darling began to use a brush rather than a pen for his work and "to whack away with any kind of tool that would make a mark approximately where I want it to go." Dissatisfied with his work and frightened by the growing atrophy in his right arm, he valiantly tried to teach himself to draw with his left hand.[6]

In the midst of the crisis in his career Jay Darling also lost his father. The elder Darling had closed his pastorate in Glencoe, Illinois, at the time John Darling was born—at the end of 1909. The Reverend and Mrs. Darling left immediately for Florida but returned in March, earlier than they had planned, because Darling's asthma attacks became insufferable. Marc Darling knew his condition was serious. From Glencoe he wrote, "The embers are beginning to burn lower—they must smoulder and go out here—but I have hopes that I may be deemed worthy, through a sincere desire to serve God in serving my fellow men, to behold a rekindled fire on another shore."

Memorial Day, 1912, Darling attended a reunion with members of his Civil War company in the church in Leon, New York, where they had enlisted a half-century earlier. His wife wrote that Marc Darling made the address dedicating the Soldiers' Monument even though it was a great tax on him.

Shortly after Mary's birth, the elder Darlings traveled to Long Island to visit "Jay and Genevieve [Penny] and the babies." They spent several weeks there, taking time out to visit the Thompsons at Glen Cove, Long Island (old friends from Elkhart, Indiana), and spent Thanksgiving evening before the open fire with Jay and Penny, John and Mary. Marc Darling died ten weeks later, February 8, 1913, at Augustana Hospital in Chicago, five days following surgery for gall-stones.[7]

Although Jay Darling had never been as close to his father as to his mother, the death of Marc Darling brought with it a troublesome emotional tug and yanked away a needed prop in the midst of Ding's uncertainty. With Marc Darling's death Jay's disgruntlement was edged in black. He was eager to escape somehow from the tinsel and glitter of New York, from the strings being tied to his conscience, and from the haunting fear that he was physically unable to give expression to the cartoonist within him.[8]

Just two weeks before Marc Darling died, Jay poured out his feelings to A. F. Allen, managing editor of the Sioux City *Journal:*

Nine out of ten men will I think say a man is a fool to work on a Sioux City *Journal* or a *Register and Leader* when he might go to Washington or New York or Philadelphia. I believe that in about eight case[s] out of the nine the advice would be wrong. . . . Nobody knows you personally except as he sees your head sticking up over your desk and nobody gives a damn whether your head stays there or another takes its place. It is a purely heartless and mercinary [*sic*] game down here. . . . The newspaper organizations down here are for all the world like a factory organization in which the foreman is responsible to the stockholders and below his level everyone is subject to the ax for the sake of dividends. . . . Besides I would rather live in Sioux City on $25 a week than in New York on $40 and would live longer and much more happily at that.

Darling's financial condition had worsened, even though he received a significant salary increase when he joined the *Globe*. He wrote to Allen:

I get, what seemed to me before I came down here, an enormous salary. I could not figure how I could spend more than half of it but at the end of the first year I had barely enough to pay my doctor bills left out of my years work. If it had not been for the money I had saved in Des Moines during the few ye[a]rs there I would be almost broke.

He was troubled by the general high cost of living in New York and the damage it had done to his financial security. He wrote bitterly that "coal and 'tail' are the only things that are cheap down here and I have [n]ot found very much saving in the former and no use for the latter."[9]

Darling's desire to escape was encouraged by some indirect inquiries from the *Register and Leader*. Editors in Des Moines had been shocked to find that Ding's cartoons had begun to appear on the front page of the rival Des Moines *Capital*, which had purchased them through the *Globe* syndicate. It was an embarrassing situation and the *Register and Leader* wanted Darling back. Darling's acquaintances in Des Moines knew he was having serious trouble with his right arm. Lafe Young, publisher of the *Capital*, wrote to encourage the cartoonist and to tell him, "I regret the difficulty you are having with your arm and hope most sincerely that your contemplated operation will be successful."[10] Regardless of his physical problems, however, both the *Capital* and the *Register and Leader* wanted Darling represented on their front pages.

Gardner Cowles and Jay Darling, with everything else they shared in common, were equally stiff-necked and proud. That made striking a bargain between them a tedious and delicate procedure. Cowles would not inquire directly about Darling's situation in New York or about his future availability. W. B. Southwell, business manager at the *Register and Leader*, apparently became a surrogate for Cowles and conducted some preliminary correspondence with Darling, assessing the situation for his publisher.

Southwell eventually expressed gratification at the possibility that Darling would "be with us again in the near future." He later wrote Darling concerning the Cowles way of doing things: "You know him well enough to know that he always maneuvers to have the other man express a wish or desire to have or do a certain thing rather than take the initiative himself." The business manager was eager to consummate the deal with Darling after having done the spadework on the agreement, but Cowles insisted on handling the final contract himself. Whether he or Cowles signed the instrument, Southwell was eager to see the job done: "I wish the meeting could take place and the deal be closed within two weeks so that we could get the benefit of your [Darling's] connection with us before our intensive circulation season is too far along."[11]

Early in 1913 Cowles was apparently not as eager to see the matter settled. The business manager reported that the publisher could not meet with Darling before April. Later, Southwell was troubled to see another Darling cartoon in the *Capital* and a news item in the same paper stating that Darling was back on the job at the *Globe* syndicate.

"Does this mean any change from your previous advices to me?"
Southwell asked.[12]

Darling, too, was struggling with his pride. He yearned desperately
to get back to the commonsense colleagues and the independence he
had seemingly left in Iowa but, as he explained to his friend A. F.
Allen, he was not eager to move too fast:

There is one more thing that I might say and that is that as soon as I can go back
to the west without seeming to retreat in the face of defeat I am going to look
up a job on some such paper as I left to come down here. My work seems to be
successful enough down here but there is absolutely no fun in it at all and if it
wasn't for people saying I didn[']t make good down in New York I would go
back to the west tomorrow.

Darling added a handwritten postscript to the typewritten letter: "I will
add this by way of encouragement. Any man who can make good on the
S.C. *Journal* in the editorial dept. can stand way above the average
down here. At least it seems so to me."[13]

Whether it seemed certain to Darling at the time he wrote or not,
he was on the brink of coming to terms with the *Register and Leader*
and Gardner Cowles. Perhaps the publisher suddenly concluded that
Darling should be on board at the *Register and Leader* in time for the
subscription campaign. Perhaps his reasons were more obscure. In any
case Cowles relented. He wrote Darling, "I will try and arrange to meet
you whenever it suits your convenience."[14] The following month, in
February 1913, Darling's cartoons returned to the familiar surroundings
of the pages of the *Register and Leader*. Ding had come home.

# 6

## *Two Homecomings*

DARLING'S RETURN to Des Moines and to the *Register and Leader* was hailed in the February 26, 1913, issue of the newspaper, which reported that the cartoonist would officially rejoin the staff effective April 1. The first Sunday after Darling returned, the newspaper scattered a collection of Ding's earlier cartoons throughout its pages. Another full page, headed "J. N. Darling, That's Ding, Comes Back," was devoted to his work.[1]

The Darlings settled into a small house on Beaver Avenue and remained there about one year. They then moved to a larger home on Ingersoll between 28th and 29th streets, where the combination of a socially active head of household and two growing youngsters occasionally caused an eruption. Darling's most uproarious recollection of life in the house on Ingersoll was of a cat with a fit that upset every dish in the kitchen, stepped in a skillet of fried potatoes and came to rest in a dish of pudding on top of the refrigerator just as dinner with company was about to be served.[2]

Ding was more comfortable both physically and emotionally. Although trouble with his arm persisted, so did he. He devised a procedure by which he could draw his cartoons with his right hand, but on a much smaller scale than in the past. His diminished arm control would not permit the broad, sweeping strokes necessary to execute the full-size originals that generally measured more than two by three feet. The miniaturized drawings could be projected on a larger surface, where a technician could then reproduce them full size. Not only had Darling found a way of coping with his troublesome right arm for the time being but he was again exercising the unbridled expression he craved and Gardner Cowles freely provided.

As might be expected, it was not long before Ding's drawings were expressing opinions contrary to those on the *Register and Leader*'s editorial page. In 1915, for example, Darling argued for national military preparedness while the editorial page condemned the world-wide arms race.[3]

His move back to the *Register and Leader* did nothing to slow the flurry of offers from some of the most prestigious newspapers in the nation. The Kansas City *Star* referred to Darling as "the best cartoonist in this broad land at present." He was encouraged to join the Saint Louis *Post-Dispatch* by Joseph Pulitzer himself, but he politely declined. He also received offers from the Chicago *Tribune* and the Chicago *Examiner*. His friend J. W. Carey wrote from Sioux City to congratulate Darling on a lucrative offer from the Chicago *Tribune* and to compliment the artist for his courage in turning it down.[4]

Darling was hitting his stride. From the peaceful surroundings of Des Moines he ventured to the metropolises, mingled with the influential and famous, and scurried back to his Iowa sanctuary to comment on events in the nation and the world. In those halcyon days prior to World War I, Darling became personally acquainted with two of the several American presidents he came to know.

Theodore Roosevelt, Darling's conservationist hero, had not run in 1908, and his hand-picked successor, William Howard Taft, had been elected president. Roosevelt, however, was so disappointed in the Taft administration that in 1912 he bolted and ran against Taft and against Democratic nominee Woodrow Wilson, who won the election. Beside a photographic portrait of himself taken in 1914 or 1915, Darling wrote in a personal scrapbook that he was at that time "at the top of my best drawing years." The photo was made when Darling joined Gardner Cowles on a trip to Washington. On that trip Darling and Cowles had lunch with Taft and met the same afternoon with Roosevelt.[5]

As world events continued their inexorable march toward world war, Darling continued to receive lucrative offers from newspapers in the nation's metropolitan areas. However, he had meant what he said upon returning to Iowa from New York City. He was not cut out to fashion cartoons according to blueprints provided by editors or publishers. Nearly a quarter-century later, Darling wrote, "I still maintain the same privileges as I have always demanded—that I will express my own honest convictions in a cartoon without fear or favor and they can run the cartoon or leave it out as they choose. . . ."[6]

Darling took pride in the record of independence he had established in Sioux City, New York, and Des Moines: "I am rather proud of the fact that never once throughout my whole period of cartooning has any editor or any influence outside of my own conscience ever dictated to me what kind of cartoons I should draw."[7]

Without question Ding had become an Iowa and a national institution and, in the bargain, a significant asset to the *Register and Leader*. Even though his cartoons occasionally took a route different from that of the editorial page, Darling devotees were legion. The daily Ding cartoon became the first item of business for *Register and Leader* readers everywhere. What did Ding have to say today? became a stock question for coffee breaks on Main Street and breakfast on the farm. Paper carriers were among Darling's most devoted fans. Even before unbundling the stacks of newspapers they were to deliver, carriers would crouch shivering in the dark, flashlight in hand, peering beneath the paper wrapping for a sneak preview of Ding's daily message. One loyal fan told Darling of a lad who walked a mile across a cornfield to see a Ding cartoon that had been saved for him. "Such stories," Darling wrote many years later, "would have been benevolent oil for the gears when the old cartoon machine was having a hard time to start."[8]

Darling wielded such influence and enjoyed such independence at the *Register and Leader* that during World War I he finally entertained an offer from the New York *Herald Tribune* to include his cartoons as part of that newspaper's syndicated service to other publications. Although the *Register and Leader* had no syndicate of its own, Gardner Cowles was discomforted by the idea that his cartoonist would consider working for another paper's syndicate while continuing to receive a paycheck from the Des Moines paper. Darling was insistent that he be given the opportunity to play before a larger audience. Cowles knew Darling was a vital asset to the *Register and Leader,* that he had been influential in the paper's steadily rising circulation, and that the *Register and Leader* could not offer the national outlet Ding demanded. One Darling advocate suggested that "Jay did more for the *Register* than Cowles did." Cowles relented and Darling, on his fortieth birthday— October 21, 1916—signed a ten-year contract with the New York *Herald Tribune* and the *Herald Tribune* syndicate.[9]

Darling had won a bloodless coup. He had broadened his horizons far more than had been possible at the New York *Globe,* he had named his own terms, and he had remained where he wanted to be. He had fully arrived as a nationally recognized political cartoonist, who was not only in demand but was available from coast to coast through a chain of *Herald Tribune* syndicate clients numbering as many as 130 major newspapers.

Even atop this crest, however, Darling was gamely trying to cope with the recurring and again worsening problem of his crippled right hand. His mother, whom he revered, died in April 1916, casting a deep shadow on his bright prospects.[10] There was also some difficulty with the publishers of the *Herald Tribune,* who were convinced that Darling's value to the syndicate was diminished by his absence from New York.

Darling yielded. He agreed to spend some of his time in New York in accordance with the wishes of the executives in charge of the powerful and respected *Herald Tribune*. According to his agreement with the paper, Darling spent several days each month in New York. He traveled there by Pullman car, often feverishly turning out three or four syndicate cartoons on the way to leave sufficient time for baseball games and socializing once he reached the city.

Ogden Reid was publisher of the *Herald Tribune*, but his wife Helen ran the newspaper. She was a liberal Republican on the same general track as William Allen White and Wendell Willkie, according to Gardner (Mike) Cowles, Jr.; and she took pleasure in arranging dinners to which industrialists, commentators, and political leaders were invited to discuss issues of the day. Darling was a regular at the dinners and through them became acquainted with such luminaries as Walter Lippmann and Bernard Baruch. Cowles was always surprised that Darling did not move to New York to stay. "He loved the monthly trips and the meetings with the great," Cowles recalled, and the New York connection gave Darling a far broader perspective in his cartooning than he would otherwise have enjoyed.

The continual traveling did become trying and tiring and kept Darling separated from his family more than he liked. Meanwhile, his right hand grew more crippled and uncontrollable. His fingers became curled and taut, and his hand took on the features of a grasping claw. Darling was conscious of the unsightliness of his hand, and he carefully hid it from view. When seated at the table, he made a habit of keeping his deformed hand beneath a napkin.

Even so, Darling remained outgoing and affable. He enjoyed the company of others and continued to make the rounds of New York social and political life. He attracted friends and admirers wherever he went. A friend recalls, "He was always helpful and sincere. And women were nuts about him. Women who didn't know him at all, had never met him, would bust right in and take him over." By comparison, the people surrounding Darling were "dull, dull, dull," says another acquaintance. "He was a human whirlwind," and those around him became colorless in his presence. "Jay was a fascinating man," says a woman who knew him for more than a half-century. "He had an attribute that I don't recall in anyone else. When he visited with you and if he liked you he seemed to be utterly enthralled. He gave you his complete attention and made you feel very witty and important which, of course, I was not." Darling was variously described as a handsome devil, a sharp dresser, and a witty conversationalist. He stood more than six feet tall, weighed about 190 pounds, and possessed a resonant bass voice.

For approximately a year in 1918 and 1919 the Darling family lived in New York, while Darling cartooned for the *Herald Tribune* syndicate and in a turnabout mailed his *Register* cartoons to Des Moines. (The *Register* dropped the *Leader* from its logotype in 1916.)[11] The family took up temporary residence in an apartment belonging to the prominent sportswriter Grantland Rice, which was on Riverside Drive near the tomb of Ulysses S. Grant. Darling was surprised to learn that his seven-year-old daughter Mary did not know who Grantland Rice was; although John, then nine years old, was well enough informed to be impressed at the company his father kept.

In the course of his continual excursions, Darling met Dr. Frederick Peterson, a neurologist then practicing in the East but a son of Sioux City, Iowa. Peterson took a personal interest in Darling and his affliction and, following detailed examination and study, determined that the malady was what Darling, the would-be doctor, had diagnosed it as being—an ulnar nerve severely damaged by the poorly mended bones in his elbow and arm.[12]

While the Darling family resided in New York City, Dr. Peterson arranged the necessary surgery on the cartoonist's elbow. The operation was an unqualified success and resulted in a new lease on the full life Darling required. Although he had devised a system for meeting his cartooning responsibilities, Darling far preferred executing his full-scale works himself. His active life in the field and on the water also was enhanced by the full use and control of his arm and hand.

Even though his second stay in New York had been brief, Darling was as disenchanted with the metropolis and the ways of its newspaper industry in 1919 as he had been when he left Manhattan behind six years earlier. He wrote, "Well, we are on our way back and I cannot tel[l] you the satisfaction I feel over the prospect of being back in a congenial field. As the well known Bard of Avon was wont to say 'I would rather live in a tent in Iowa than to be a door keeper in these houses of wickedness forever at three times the salary.' "[13]

The *Herald Tribune* had apparently been paying a premium for the privilege of having Darling close at hand. In any event, Darling was again returning to Iowa and he was again paying a price for his decision: "I am making something of a financial sacrifice for the sake of the pleasure of working with a free hand and being my own editor in chief. . . . The people of Iowa think more to the square inch than the people of New York think to the square mile and if it is in me to do a man[']s work in the cartoon game I can do it in Iowa and I never could in New York.''[14]

Although opportunities continued to cross Ding's path, the Darling family had made its last move to the "big city" and back home to Des Moines. Jay Darling's life would be full and his talents in demand in several quarters—including the office of the president of the United States—but Darling never again would uproot his family or consider any place but Des Moines and Iowa his home. When he returned in 1919, it was with a determination to sink his roots in the fertile, black, central Iowa soil. Even though he was a man who belonged to the nation, his deepest loyalties and his lifelong commitments were to the values he had learned in Iowa. He stood on the threshold of his golden years, during which he would be thrust into the national limelight and asked to perform a nearly impossible task for conservation and his country, his political independence would be tested, his name would be nearly as familiar as the president's, and his children would grow to maturity.

# 7

## *Good Times*

SOON AFTER Jay Darling and his family returned to Des Moines, the now famous cartoonist first met Herbert Hoover—a fellow Iowan, a future president of the United States, and a man who would become a durable friend. Hoover's name had become a household word in the United States and abroad thanks to his skillful organization of food relief operations during World War I. A native of West Branch, Iowa, and about two years older than Darling, he was an 1895 graduate of Stanford University and a mining engineer, prospector, businessman, and wealthy Quaker humanitarian.

Hoover was a cartoonist's curse. Darling had executed likenesses of the future president even before the United States became involved in World War I but was invariably dissatisfied with the results. Hoover, unfortunately for political cartoonists, "had the most average-looking face," Darling lamented. In 1919, when Hoover made a visit to Des Moines, Darling visited him to study his features in the flesh. Darling introduced himself to Hoover. "Oh, the cartoonist?" Hoover inquired. The food relief organizer thanked Darling for the cartoons in support of his international food distribution project. Darling studied his subject's face and made some rough sketches for future reference. It was the beginning of a warm friendship between the Hoovers and the Darling family.

Also in 1919 Darling quickly drew what was to become the most popular and most reproduced cartoon of his career. He was hard at work behind the locked door of his *Register* studio, when word of Theodore Roosevelt's death reached the newsroom. The editors pounded frantically on the studio door to inform Ding of the death of his friend and fellow conservationist. Darling's deadline was near. He dashed off a quick sketch for that day's paper and intended to do a more refined illustration later for the syndicate. His "Long, Long Trail" became such an immediate favorite that a second version was never attempted. The sketch has been reproduced more than any other Darling cartoon—in granite, copper, wood, and concrete as well as on paper, and it is displayed in hundreds of public places.[1]

Darling's financial condition and his national reputation were secure. Assured of a solid foundation, he began to invest his time and skills in a variety of new experiences. He plunged into a kaleidoscopic range of interests that gave him pleasure, underlined his intense drive to enjoy life, rounded his international knowledge, and led him irresistibly to a second career devoted to conservation of natural resources.

He became one of the first Iowans to obtain a flying license.[2] He would don his aviator outfit, complete with goggles and leather helmet, and take a flight for the pure joy of it. He relived his pony-riding days, but in the more refined atmosphere afforded by the Des Moines Wakonda Saddle Club. Darling, a leader in the club and one of its presidents, interested his children in riding. John and Mary participated in horse shows as they grew older and were successful competitors. Other members of the Wakonda Saddle Club included Mr. and Mrs. Gardner Cowles; Mr. and Mrs. Vernon L. Clark; and several members of the Hubbell family, Des Moines pioneers and influential in the banking and insurance businesses in the city. He also devoted large amounts of his time and talents to city recreation programs and to other local projects such as day care centers for the children of working mothers.[3]

Darling was instrumental in organizing the Iowa division of the Izaak Walton League of America, one of the most progressive and industrious of the League's chapters.[4] Darling recalled some of the trials of conservationists before the organization of the Izaak Walton League:

Such remedial efforts as we could conceive at that time were wholly limited to personal appeals and attempts to convert a legislator here and there in the hope that he might carry the message to the State Legislature. Such personal appeals without any organized backing got nowhere. The only organization[s] that showed above the horizon in those years were scattered groups of hunters, here and there, who were bound together in the interests of a more or less private hunting club, and they were scarce.

The news of Will Dilg and his projected organization of the first Izaak Walton League to support a project on the Mississippi River near Lansing, Iowa, was "welcome news," wrote Darling, "to our little group of sportsmen . . . who had watched the Prairie Chicken disappear, the flight of migratory waterfowl dwindle from clouds in the sky to sporadic flights of remnants, pollution, siltation and complete disregard of fishing laws, and the Will Dilg idea gave us the first encouraging hint of how to proceed.'' Darling also credited the faculty at Iowa State College and the State University of Iowa with lending their prestige and scientific leadership to the early efforts to organize chapters in the League.[5] Several decades later Darling wrote, ''I get quite a chuckle now and then when I read about the enthusiasm and success of chapters in some of the towns where long ago we tried to set up chapters and got thrown out on our ears.''[6]

For all his activity, however, Darling was not a well man. He suffered continually from heart problems and from chronic bronchial asthma, an ailment similar to the condition that had plagued his father. In 1921 he was obliged to make a quick trip to the Mayo Clinic in Rochester, Minnesota, ''where they do things very thoroughly and in a hurry,'' to find an infection that was the source of much difficulty and pain. He later was hospitalized with a malady, probably a bronchial attack, that left him very weak and very nervous, too ill to receive visitors, and in a hospital bed for nearly a month.[7]

Although his health was uncertain, Darling's career was on a steady upward path. In 1923 he drew a four-panel cartoon lauding hard work, dedication to others, and devotion to personal goals. The first panel showed an ''orphan at 8'' who ''is now one of the world's greatest mining engineers,'' the second presented the ''son of a plasterer'' who had become ''the world's greatest neurologist,'' the third was devoted to a ''printer's apprentice'' who had become president of the United States. The fourth panel suggested, ''But they didn't get there by hanging around the corner drug store.'' The cartoon, titled, ''In Good Old U.S.A.,'' won the Pulitzer Prize for cartoons in 1924. The personalities illustrated in the cartoon were Herbert Hoover, Dr. Frederick Peterson, and Warren G. Harding. Ding's prize was the second Pulitzer given for cartoons.[8] The first went to Rollin Kirby of the New York *World* in 1922; no award was made in 1923.

The Darlings were enjoying Ding's fame and affluence in a unique home on Terrace Road, immediately south of Terrace Hill, the Victorian mansion owned by the Hubbell family. The Terrace Road home was painted brown, according to Darling's taste, just as his other two Des Moines houses had been. The home was creatively constructed on several levels and, as the Darlings' tenure there lengthened, it gradually became an intricate marvel of understated elegance on a total of five levels. Because the building was constructed overlooking the Raccoon River Valley, its size was not apparent to the casual passerby on the street. Darling toiled to construct a separate studio beside the house on a lower level. Only the tall chimney of the multistoried studio was visible from the street level. A unique Darling-designed swimming pool was not visible from the street at all. The home and its appurtenances were objects of much attention in Des Moines and in the nation and were repeatedly the subjects of feature articles in such nationally circulated publications as *Better Homes and Gardens* and *House Beautiful*.[9]

The Darlings also took up travel. Jay was in far better financial condition than his father had been, and he was just as determined to expose his children to international experiences and perspectives. In 1924 the family crisscrossed Europe. Darling spent nearly two months behind the steering wheel of an automobile as he toured France and Italy, taking time to meet the inhabitants along the way. Penny and the children had been in Germany, and Darling later met them there to continue the European sojourn. Foreign travel and study became commonplace for the Darlings in the 1920s. In 1922 John, at the age of twelve, and Mary, at the age of ten, went to Switzerland for a year's study at Lausanne. Penny meanwhile toured Europe and Africa.[10]

Darling believed firmly in self-reliance and sought to instill in his offspring the values of independence and intellectual enterprise. From very early in life the children were left to fend largely for themselves as part of Darling's scheme for developing their resourcefulness and self-sufficiency. Jay found his alter ego in his son. John was a bright, healthy lad who possessed several of his father's attributes. Although Jay had been dissuaded from a career in medicine, he determined early on that his son would not be denied that opportunity. Besides a strong physical resemblance, John exhibited some of his father's musical talents. At about eleven years of age, he became a member of the choir at St. Paul's Episcopal Church in Des Moines, reconstructing an event that had been important in Jay's own maturation. John also took piano lessons and became skilled at the keyboard.[11]

Jay Darling was a strict and stern father. In addition to leaving his children largely on their own while he and Penny traveled the world or looked after business, he held John and Mary to rigid standards of behavior. Jay Darling, the preacher's son, had set his own course and had developed his own personal standards, some of which were at variance with the teachings of his parents. He was his own man and easily sloughed off the straitlaced puritanical standards by which the Reverend and Mrs. Marc Darling had tried to guide him. He was a personable character and an accomplished social mixer. He drank socially but rarely to excess. In his long lifetime he smoked tobacco in the form of cigarettes, pipes, and cigars, with total disregard for the asthmatic condition that repeatedly threatened to kill him.

Even though he had set his own standards, he held his son and daughter to virtually the same code Marc Darling had set out for his sons Frank and Jay. Darling could swear with color and skill, but he reserved his use of strong language for occasions that warranted its special impact. He would not, however, abide the most innocent slip of the exasperated tongue of his son or daughter. On a train trip to New York, Mary and Penny were working on a Christmas list. As they neared the end of their task, Mary suddenly realized they had forgotten the name of her godmother. "Oh, my Godmother!" she exclaimed. Darling bounded into their drawing room from an adjacent compartment. "Don't you ever speak to your Mother like that!" he boomed.

Darling was "firm and very careful" with his children. A friend who watched John and Mary grow up suggested that Darling was "a pretty tough father. He was tough on Mary and he was tough on John, too." On one occasion, Darling was especially quiet and indifferent toward Mary. Jay had heard from a neighbor that Mary had ridden in the street in violation of her father's orders. His chilly attitude was painful enough, but Mary was particularly hurt to think her father would believe a neighbor without asking her for details or confronting her with his evidence.

John and Mary also feared disappointing their father, who not only claimed that anyone with moderate intelligence could become an expert in virtually any field but who continually proved his point by his personal example.

Darling was a perfectionist in his work and his life, with all the difficulties, despair, and frustration the quest for the impossible can create. He held in highest regard those who shared his rarified objectives. His occasional temperamental outbursts were generally aimed at persons whose performance showed a lack of industry, integrity, intellectual effort, or faultless intuition.[12] He could be difficult to work for; yet his employees without exception expressed reverence for their colorful employer. His outrage was usually short-lived as well. A longtime friend recalls a Darling explosion caused by the friend's failure to follow through on an order issued by the cartoonist. When Darling learned he had never actually made the assignment, he was painfully and deeply embarrassed. His agony nearly brought him to tears, the friend recalls. Darling could erupt for more prosaic reasons as well. Once, when a car in front of him stopped abruptly, Darling's car bumped into it. Jay got out of his car, went up to the other driver, and "rapped him one" right through the open window.

Ding also laughed readily and enjoyed the vagaries of human existence, traits that served him well as a political cartoonist. His challenge was to digest voluminous information concerning the world about him—by thoroughly reading at least six newspapers each day—then to decide which issue to comment upon, and finally to synthesize his comment in a metaphorical, simple-to-understand illustration.[13] Darling drew upon his knowledge of biology and the Bible, his interest in medicine, his love of the out-of-doors, and his keen sense of observation and contrast in constructing his illustrated messages. Not surprisingly, Darling's writing and speaking also demonstrated his metaphorical skills.

Those skills had never been more appreciated by Ding's hundreds of thousands of devotees throughout the United States nor was his humor ever more necessary than in the troubled Midwest of the 1920s, where the Great Depression was preceded by a severe agricultural crash. Yet at the very height of his fame, fortune, and demand Darling nearly came to his end.

# 8

## *Taste of Doom*

IN MARCH 1925 Jay Darling was stricken with peritonitis, and for several weeks it appeared he would not survive the attack. The New York *World* reported that Ding was "at the brink of death" and that his doctors "hold out little." The *World* also expressed a less than confident wish for Darling's recovery. In responsible newspaper fashion, reporters and editors scurried about gathering the details of Darling's active life and compiling detailed obituaries. The stories were set in type, galley proofs were made and corrected, and in newspaper backshops across the nation the obituaries were marked "Hold," pending official word of the cartoonist's death. Presidential Aide Carter Field drafted a statement for President Calvin Coolidge, for use in the event of Darling's seemingly imminent end. The statement was approved by Coolidge, but only after he had added the following lines in his own hand: "I feel his loss because he was a personal acquaintance. The nation feels it because of his public services."[1]

Other prominent persons, including U.S. Secretary of Commerce Herbert Hoover, expressed their condolences. Hoover said of Darling: "His insight into national life lifted his cartoons into the high rank of great and trenchant editorials. His kindliness and humor were but the reflections of his own character."[2] Members of Iowa's congressional delegation expressed their statements of grief for inclusion in the obituary.

Somehow, the news trigger was mistakenly pulled, and wire service stories reported Ding's death in papers around the country. Several prominent editors wrote their own laudatory eulogies. And Darling had the pleasure of laughing as he read them all while still confined to his hospital bed. Clem F. Kimball, president of the Iowa senate, and Walter H. Beam, senate secretary, sent a get-well resolution from the body of lawmakers. Darling later claimed that greeting, endorsed even by senators who had felt the point of his pen, marked the beginning of his recovery.[3]

Darling remained a sick man, too ill to draw, for approximately a year, during which time several honors were accorded him. His alma mater, Beloit College, allowed him "sweet retribution" when it conferred the honorary degree of Doctor of Letters upon the once errant art editor of the *Codex*. Ding's fame was no secret to the faculty at Beloit, and some members had decided that the fun-loving bass, the minister's son from Sioux City, the tardy and occasionally disruptive hell-raiser had turned out to be a constructive citizen after all. The dignity of those remaining faculty members who had been offended by Darling's sketches had healed in the intervening quarter-century. Some of his classmates, including the editor of the collector's item 1899 *Codex,* had become trustees of the school. The Beloit chapter of Phi Beta Kappa also elected Darling, the man who claimed he flunked "damn near every course" during one year, to honorary membership in the scholastic society. The stack of congratulatory letters Darling received following the award of the honorary Litt.D. degree included one from the famous magician Houdini.[4]

Even as Darling lay confined to his bed, his fame was confirmed in another unusual way. A Saint Paul, Minnesota, newspaper reported that a man claiming to be Ding Darling had inveigled a free meal at a local restaurant. Had the victims known the cartoonist by more than the name "Ding," they would have known trading on his name was not in character. In any case, Darling had been seriously ill in Des Moines for more than a month when the incident occurred.[5]

Although Darling was incapable of drawing for more than a year, he was not inactive. He cast about for other means of investing his resources and other skills, in the event his cartooning days were at an end. In November 1925 he inquired into the possibility that the Sioux City *Journal,* where his cartooning career began, might be up for sale because of the serious illness of one of its principals. "If that is true," Darling wrote, "then there can be no impropriety in suggesting that I (with Gardner Cowles back of me, of course) would like to be considered a possible purchaser." Darling's informant replied that the person in question had made a remarkable comeback from pernicious anemia and that the Perkins family was not disposed to put the paper on the market. Darling was disappointed: "Ever since I left the *Journal* almost twenty years ago, I have harbored an ambition to live some of those days over, and if opportunity ever offers to try it I think I will make the attempt." And Darling, ever the frustrated medical man, offered his unsolicited opinion that the reported recovery was contradictory to all medical experience and theory.[6]

When Darling's cartoons returned to the front page of the New York *Herald Tribune,* he was welcomed back in an accompanying article written by William Allen White, the renowned Emporia, Kansas, newspaper editor.[7] The Milwaukee *Journal* editorialized:

Thanks to a kindly Providence that cares alike for the man of one talent and the man of ten talents, J. N. Darling comes back to us the same Ding he was when he falteringly laid down his pen a year ago. For twenty years he had looked from the side of the road at the passing parade of humanity, sketching in bold strokes their foibles and their virtues. He had been an interpreter of America to America. Thousands had enjoyed and profited by his cartoons. Then an invisible hand tapped him on the shoulder and he took up his easel and departed. The way led down into a valley that is as old as humanity, and there Ding was left very near to the end of his journey. Out of that valley he fought his way back, and he is again by the side of the road, with eyes that see a little deeper than most men and a hand to make us see what he sees. . . . And while Ding has been gone, there has been a place unfilled in that unique department of American journalism—the cartoon. Others drew on, but Darling alone knew the secrets of his own canvas. . . . Welcome back, Ding.[8]

Darling did see a little deeper than most men following his face-to-face encounter with his own vulnerability. The result, however, was not caution for his physical well-being but a renewed determination to squeeze into the life allotted to him the fullest possible measure of living. A year away from his drawing board had affected his self-confidence as well. When he received a letter at the end of his hiatus complimenting him on his work, he responded appreciatively: "I had just started in drawing again with many grave qualms and misgivings and I had a weird sinking feeling in the pit of my stomach every time I turned one of those first cartoons in."[9]

Darling and his family had made a dark passage, had tasted a lingering, drawn-out grey year of doom. But Ding again prevailed. He was busy at his drawing board and the gloom was gone. Shortly after his return to the nation's front pages, Darling was awarded his second honorary doctorate—a Doctor of Laws degree presented by Drake University in Des Moines. With the return of his vigor, Darling went back to the lucrative business of informing and entertaining an appreciative national audience. He was receiving an annual salary of $26,000 in 1925. In addition, he received a fifty percent commission on syndicated sales of his work—in excess of $11,700 per quarter. His total income came to approximately $100,000 each year for many years. He later received additional income from stock he owned in the Register and Tribune Company, in a Des Moines insurance company, and in other enterprises.[10]

Darling had hardly escaped the threat of peritonitis when he was cut down by influenza that developed into pneumonia and kept him low for four weeks. With such difficulties and distractions his cartooning became a chore, and he suggested that ''now I conduct my business like an old horse hitched to a milk wagon. It seems like routine.''[11]

In 1926 Darling's original ten-year contract with the *Herald Tribune* syndicate expired. The late Agnes McDonald, longtime secretary to Gardner Cowles, recounted how the Darling-Cowles friendship met a strenuous test in the midst of negotiating a new contract. Since Darling had first made his agreement with the *Herald Tribune* in 1916, the *Register* had organized its own syndication service. Gardner Cowles assumed that when Darling's New York agreement ran out the cartoonist would sign on with the *Register* syndicate. Cowles, in fact, insisted that Darling become a member of the *Register* syndicate team. Darling replied to the effect that he saw no reason to make the shift. He argued that he owed the *Register* nothing—that, in fact, he had helped build the *Register*'s circulation. Darling said he had no complaints with the way the *Herald Tribune* had treated him and intended to renew his agreement with the New York paper.

The two reportedly ''got mad and cussed at each other.'' For days Darling and Cowles sulked and fumed and spoke only intermittently. Meanwhile, the question of the contract remained in limbo. In the end, Darling—strong-willed, stubborn, independent, and in great demand—had his way. He renewed his contract with the *Herald Tribune*.

The Darlings needed a change of scene. They again packed their bags and struck out for a part of the world they had not fully investigated. In 1927 they visited Cuba and Mexico, including Yucatan. Darling made a busman's holiday of it by keeping a journal of the trip to Cuba, complete with his original detailed illustrations.[12]

John and Mary, meanwhile, were prominent in the Des Moines and Iowa society news section of the *Register,* whether pictured astride their horses or visiting their parents during school breaks. Mary attended Emma Willard School in Troy, New York, and John was a student at Princeton University. John had chosen Princeton ''because of its natural science departments in which it excells [sic]. . . .''[13]

Darling plunged again into a variety of activities in his town and in the process had a hand in the creation of the Men's Garden Clubs of America, headquartered in Des Moines. Fae Huttenlocher recalled that when she was elected president of the Des Moines Garden Club she inherited a treasury that had been depleted at the request of Jay Darling. The cartoonist had managed to persuade the club to donate $400 to pay a landscaping firm for a planting plan for Greenwood Park in Des Moines. Mrs. Huttenlocher's first order of business, insisted upon by several irate members of her board of directors, was to get the $400 back.

When she asked Darling to return the money, he refused. "It is the first worthwhile thing your club has done," he claimed. "What are you anyway, just a bunch of petunia growers? Why don't you get out and make some money like other organizations?" he asked. That was in the autumn of 1929. In response Mrs. Huttenlocher organized a flower show and enlisted Darling to head up the men's division. The show was a rousing success, and with great pride the Garden Club donated profits of more than $4,000 to several Greenwood Park projects.[14]

The Des Moines Men's Garden Club was subsequently organized in early 1930, with Darling as its first president. At a meeting of similar clubs in September 1932, the Men's Garden Clubs of America was officially organized and Harold Parnham of Des Moines, a friend of Darling's, was elected one of the organization's first directors. Ding had made it possible financially for Parnham to attend the organizational meeting. Parnham, who in 1977 was the only surviving member of the original board of the Men's Garden Clubs of America, first met Ding when the cartoonist became a customer at Parnham's nursery and garden care center. Parnham recalled grading and reseeding Darling's new lawn and installing a new crushed rock driveway and the next year the best grass grew in the rock driveway and none on the proposed lawn area.

Jay maintained his correspondence with Herbert Hoover and occasionally visited him at Rapidan Camp in the Blue Ridge Mountains for several days at a time. Two years later when Hoover was campaigning for president, Darling expressed his respect for the candidate's values and his motives: "Hoover is a real friend of man and whatever his attitude on the economics of the McNary-Haugen bill [an early effort to subsidize farm prices through government supports] is he can be depended upon to look carefully into the welfare of the human race. How in the world he ever got tagged as a spokesman for big business is a mystery." Darling was on hand for the Republican National Convention in Kansas City in 1928 to see his friend Hoover nominated for the presidency at the end of a sleepless week.[15]

Ding was a keen observer of the political scene, and he had commented on the foolishness and the forces of American politics and politicians since the McKinley-Bryan race in 1900. He was fascinated by the political process, and he understood it well; yet he distrusted the tendencies of government bureaucracy as much as his father had distrusted the bureaucrats of the mid-nineteenth century. From a distance he could be detached and could comment objectively on the accomplishments and mistakes of both major political parties in the first quarter of the twentieth century. Although his own views were well known and frequently expressed in his cartoons, Darling treated personalities of both parties in a relatively gentle, evenhanded way.

He preferred to fashion his cartoons in an atmosphere of privacy and calm and rarely drew when angry. He often left his most critical and cutting cartoons to "cool" overnight and nearly as often threw them away the following day. As one result, the great majority of his drawings exhibited what Darling's fellow cartoonist, Rube Goldberg, a quarter-century later termed "Ding's gentle humor." Goldberg once told Walt Kelly, the creator of the "Pogo" comic strip, that Darling had "developed the art of gentle ridicule better than anyone in the world. There is no defense against it."

Politics had been a subject of deep interest to Darling, and his detached and incisive observations of the political process had been influential in his wide acceptance and phenomenal success. But Ding's political independence was about to be seriously challenged.

# 9

## Political Possibilities

JAY DARLING had become a revered member of the Des Moines *Register* and *Tribune* staff. He was officed in a corner room on the eleventh floor of the Register and Tribune Building, in downtown Des Moines, several floors above the newsroom. In 1931 Merle Houts, Darling's personal secretary, first took up her post outside Darling's enclosed studio-office. Her responsibilities varied but generally did not include answering Ding's telephone. He did that himself. Her career lasted until she closed Darling's office following his death thirty-one years later.

Darling occasionally strode through the newsroom, and when he did heads bobbed up from their work. One staff member recalled, "I remember how dynamic and forceful he would seem as he would stride through the newsroom, sleeves of his always-white shirt rolled up and his vest neatly buttoned . . . men usually wore them in an office back in those days." He was a legendary figure at the R & T, and veterans took pride in casually pointing out the famous Ding to the uninitiated.

Visitors to Darling's lair ran the gamut from an outspoken local barber[1] to comedian Will Rogers to one of the "under of the underlings" on the newspaper staff. One such visitor was Elizabeth Clarkson Zwart, who recalled Darling's studio as a wonderful place, with great corner windows. Ding, as always, was seated at his drawing board. He never used a desk in his *Register* office, preferring to do his business as well as his cartooning at the large board. He explained to Mrs. Zwart, as he began to draw in the lower left portion of the paper on the board, that a baby's face "climbs up its head." He continued to draw, finally completing several sketches of a baby's head in progression to illustrate his point. "When he finished," Mrs. Zwart lamented, "I walked out. I'll never know why I didn't reach over and pick up the drawing and take it with me."

Rodney Fox, who had been a young reporter at the *Register,* and went on to teach for nearly forty years as a professor of journalism at Iowa State University, remembered the anticipation that accompanied Darling's daily contribution to the front page. "When the cartoons came down, everyone wanted to see what Ding had to say for that day," he said. "Sometimes they had a hell of a time getting the drawing down to the engravers."

Highlights of 1931 included a Rotterdam cruise for Jay and Penny; Mary's graduation from Emma Willard School[2]; and through the Iowa Division of the Izaak Walton League, Jay's success in having the 1931 Iowa General Assembly establish a nonpolitical state Fish and Game Commission of five members. Darling was subsequently named one of the commission's original members.[3]

Darling, unquestionably established as one of the nation's leading cartoonists, turned his avocation as an outdoorsman to the serious business of conservation of precious natural resources. As an observer of conservation policies in Iowa and elsewhere, he had been distressed at the interference of politics in the life-and-death business of conserving limited resources. Officials in charge of saving irreplaceable water, air, and land were hog-tied. Because of short-term political considerations, they could not institute what seemed to be only moderate and sensible programs for the long-term maintenance of America's natural bounty. Darling's love of nature and his distrust of politicians combined to create a fiery desire to see the politician's hands removed from the natural resources cookie jar.

While Darling was concentrating on politics in conservation, he was also keeping watch on the White House where his friend Herbert Hoover resided as president and where the outlook for Hoover and the Republican administration was growing continually bleaker. The 1929 crash had upset the economic applecart and Hoover, diligent and determined as he was, could not retrieve the fruit and right the cart. His political days were numbered.

Darling continued to visit his friend and to offer what encouragement he could. Ding had always appreciated the fact that Hoover had invited the Darling family to visit him and Mrs. Hoover at the White House in the golden days shortly after the inauguration. Following that visit Mrs. Hoover wrote, "Along with happy memories to Mr. and Mrs. Ding Darling and the two dinglings." As his part of a pact with Mrs. Hoover, Darling added a small "x" to his signature on the cartoons he personally completed. It was a private code he used to let Mrs. Hoover distinguish between the pure Ding cartoons and those he roughed out in pencil to be finished in ink by his assistant, Tom Carlisle. Following

Mrs. Hoover's death in 1944 Darling forgetfully added the ''x'' to some of his cartoons out of habit.[4]

Jay also recalled an especially gratifying visit with the president on a day when the nation's chief executive and his favorite political cartoonist played ''hookey.'' ''Herbie'' Hoover had told Ding that the only time the president could be alone was when he was praying or fishing, and Hoover hardly had time to do either anymore with eight Secret Service men and the White House physician always nearby. Darling described the incident with undisguised glee:

One day when he and I were out horse back riding in the Smoky Mountains we plotted to duck away from the cohort of body guards that usually rode ahead and behind and sneaked off on a side trail that was hardly visible and rode like hell until we felt pretty safe. Then we took another side trail and wound up at an old deserted observation tower of the Forestry Service. We hitched our horses off in the brush[,] climbed up into the top of the tower and spent the whole day there—cooked our own lunch, built a dam in a little creek that ran near the tower, caught a mess of trout for the frying pan, but the most fun was watching from the top of the tower the secret service men scouring the countryside to find the President. It is the only time in my life I ever saw that man really happy and unrestrained.[5]

Prohibition was the talk of the town and of the country. One exchange of correspondence between Darling and Hoover referred to Prohibition in a light vein. It also underlined Darling's propensity to ''practice'' medicine and provided a glimpse of the Hoover humor so many thought did not exist. Darling had just returned from a visit to the Mayo Clinic, where he made it a practice to talk medical shop with the professional staff. Tongue in cheek, he reported to President Hoover:

Dr. Will Mayo told me he had discovered the only real indictment against the prohibition law. The indictment was based on the fact that there are no longer sufficient cadavers to supply the medical clinics and furnish the medical students with laboratory materials. Prohibition has eliminated the derelict and has caused such general prosperity that friends are always able to provide proper burial. [Deceased and unclaimed derelicts in the past had been used by medical schools and then interred in paupers' graves.][6]

Hoover, besieged on all sides for a solution to the nation's desperate ills and surrounded by detractors, replied with his own brand of dark humor: ''I deeply regret the shortage of cadavers. If they were to give me a little more authority I would supply this deficit.''[7]

Early in his administration, Hoover appointed a special citizen's committee to study federal lands and to recommend a program of management and allocation. At that time, according to Darling, there were more than 200 million acres in federal lands, a good deal of which had been abused by the grazing interests. No agency of the federal government had exercised jurisdiction in any coordinated or organized way over the government-owned areas. Gardner Cowles, Darling's publisher, was appointed a member of Hoover's committee. After careful study the committee recommended that any areas of potential value be given to the states in which they were located. That left approximately fifty million acres with no value except for recreation at some future time. Those acres, embracing largely inaccessible and mountainous terrain or unproductive arid soil, were classified as wilderness areas.[8]

Darling's side-of-the-road political stance was given a jolt in 1931 when John W. Carey, a former colleague on the Sioux City *Journal*, suggested in his *Journal* column, "The Rear Seat," that Ding should run for the U.S. Senate.[9] It was no secret that Darling was a conservative Republican. His GOP supporters were eager to see so famous and respected a figure as Ding take on the incumbent Senator Smith W. Brookhart, a native of Missouri and a resident of Washington, Iowa. Brookhart was also running as a Republican, but his views were progressive and described as being similar to those held by the better known Senator George Norris of Nebraska.

Darling swung almost immediately into action, disavowing any interest in running for the Senate, even though Republicans with their fingers on Iowa's pulse were claiming that Ding was the only Republican who stood a chance of being elected in Iowa in 1932. A Marshalltown *Times-Republican* editorial quoted Darling as having thanked the Sioux City *Journal* for "kindly intentions." The editorial further quoted the cartoonist's plea to be left out of the race. "I have been discreet in both my public and private life," Darling insisted. He said he did not know why "anyone should lay the orphan on" the Darling doorstep. Ding, the editorial concluded, couldn't help but look on the proposition as a huge joke. The rumors were spreading, however, and Darling discovered that quieting them was like trying to capture flies in a jar. The Des Moines *Plain Talk* referred to the "Darling movement." The Sioux City *Journal* again entered the act. Following up on Ding's remark suggesting the nomination idea was a joke, the *Journal* said, "We trust that, in view of the widespread interest in the Davenport *Times'* suggestion that he be a candidate for United States senator, Mr. Darling will not take exception to our reprinting his letter for general enlightenment as to his attitude."[10]

The Kossuth County *Advance* also credited the *Times* with having originated the suggestion. Whether the idea came first from the *Journal* or the *Times*, Darling was having no luck quashing it. The exasperated cartoonist was quoted in the Storm Lake *Pilot-Tribune*, fully nine months after the "Ding for Senator" campaign had erupted: "I haven't any political hat to throw in the ring and I haven't any relatives who want political jobs. Whatever I may say about Brookhart is born of my dislike of political quackery, and not because I have any political ambitions of my own. No, I don't want to be a senator; but I'd be very grateful for a chance to vote for a good one."[11]

In 1932 Darling was a delegate to the Republican National Convention in Chicago and an extraordinary delegate at that. Governor Dan Turner of Iowa, also a member of the Iowa delegation, was left off the national Resolutions Committee in favor of Jay Darling, possibly at Hoover's request.[12] Iowa's Republican contingent at the GOP convention included persuasive elements convinced that the state had grown weary of Prohibition and that the Iowa electorate was teetering and about to fall off the wagon. Turner was a professed "dry," but Darling supporters argued that his cartoons provided ample evidence of his "wetness." Darling's friends also argued that Ding was the state's most outstanding citizen and that he could wield more influence for Iowa in the Resolutions Committee than could any other representative of the Hawkeye State.

The election of Darling was accomplished, but only following approximately two hours of heated discussion. That done, Darling broke from the crowd of admirers, "many of whom were women," to talk with a reporter. He said that he had not cared to have the honor but believed he could help get Iowa "some of the things of which it stands in sore need." It was also common knowledge at the convention that Darling was a personal friend of President Hoover and a frequent guest at the White House.[13]

The Sioux City *Journal*, referring to the gaggle of women surrounding Darling at the Chicago convention, commented:

No one knows better than Mr. Davis [the reporter] that Mr. Darling simply is running true to form when he has faire ladyes buzzing around him. Back in 1900 Mr. Davis was the Sioux City *Journal*'s star reporter and Mr. Darling was its cub. Even then he was a favorite with the fair sex. Possibly it was because he had a deep bass voice and knew how to play a mandolin and could sing sweet serenades in the moonlight.[14]

Darling's position on Prohibition was moderate and not necessarily "wet." A decade later Ding wrote to an advocate of temperance: "All you say of the evils of liquor traffic is true and if you can remember that far back you will recall that in the previous prohibition campaign I fought on the side of prohibition until the mess became too rotten and corrupt for further toleration.'' Darling also said he agreed with President Hoover who contended that total prohibition had been instituted before the antialcohol educational program had been allowed to run its course and accomplish its purpose.[15] Ding reiterated the point in another letter: ''We did lose the splendid advancement toward temperance which had been accomplished by education progress when we clamped on total prohibition.'' A few years later when he became head of the National Wildlife Restoration Program, Darling was faced with the choice of total prohibition of shooting or rigid restrictions. He chose the latter, ''on the lessons which total alcoholic prohibition had taught us in law enforcement.''[16]

With Darling taking such a conspicuous part in the Republican convention, the wishful thinkers continued to stump for his Senate candidacy. The Storm Lake *Pilot-Tribune* commented: ''Although he politely but firmly refused to consider himself a candidate for the United States senate when his name was proposed some months ago, there are evidences that the political bee is buzzing around under the hat of J. N. Darling.'' One of those evidences was that, in addition to supplanting Iowa's governor on the Resolutions Committee, Darling had taken center stage to make the nominating speech for Hanford MacNider of Mason City, Iowa, for the vice-presidential candidacy. At that time he defied custom by immediately identifying the nominee and removing any suspense that may have lingered among his listeners. The move was not intended as an oratorical ploy but was made to avoid embarrassment for Darling. The previous evening he had exceeded his allotted time as he made a speech on the Prohibition plank. He was unceremoniously ordered to his seat before he had revealed for which side of the issue he spoke![17]

Darling understood the political process, just as he understood human nature, and he took pleasure in poking and jabbing at it and occasionally influencing it as a citizen-politician and as a respected commentator. In elected political office, however, he saw only threats to his precious independence as well as entrapment, compromise, and frustration. He enjoyed being a free spirit and believed that he could realize greater accomplishments through his uninhibited and unrestrained nationally circulated comments and criticisms than he could from behind a senator's desk. Election to the Senate would bring his cartooning career to a halt for a minimum of six years and would drastically reduce his income for at least that long.

Darling was willing to work actively and contribute unstintingly to his Republican interests through the GOP machinery, and he did so for many years. He viewed elected political office, though, as a personal and professional tomb. The suggestion was flattering. It had caused a tremendous stir for months in the newspapers of Iowa.[18] It had made his name even better known in the state; and it had shown the candidates for national office, including Franklin Delano Roosevelt, that Jay Darling had a devoted following for whatever banner he chose to carry.

*An Album*

BY ANY MEASURE Jay Norwood Darling was one of the best cartoonists in the United States as well as one of the best known. He was also one of the most durable. He began his cartooning career in 1900 and retired from it nearly fifty years later. His life began soon after Custer's Last Stand and ended soon after the Berlin Wall. Darling witnessed staggering changes in the world about him. With his drawing instruments he brought clarity and, much of the time, humor to the perplexities of American life throughout the first half of the twentieth century.

In the opinion of Richard West, a student of cartooning history, Ding was in a class by himself; he developed his own style of cartooning and had no rivals, good imitators, or successors.

Just as any artist labors to perform with beguiling ease, Darling sweated over his drawing board to make his cartoons appear simple. In 1936 he wrote that "those few lines which finally appear in my

*This photograph of Jay N. "Ding" Darling was taken by Herbert Schwartz, a Des Moines* Register *and* Tribune *photographer, and appeared in the* Tribune *June 25, 1942.*

pictures I only arrive at after hours and hours of effort to simplify. . . ." He often said the hardest part of cartooning was persuading anyone that it was hard work.

Darling was zealous in his attention to detail. He took seriously any criticism that improved the accuracy of his work. In 1940 he told a writer for the *Saturday Evening Post,* "I was rebuked for putting teeth in the mouth of a vegetarian whale. Carpenters write in when I leave floor girders unsupported, and if I have the sparks flying the wrong way from a grindstone, the whole machinists' union protests." Darling thanked his constructive critics, including the head of a rope company who taught him that fishline and twine are twisted in the opposite direction from the cordage used on ships.

In turn, Ding's advice was continually sought by and patiently given to scores of would-be cartoonists and their parents. He wrote of one hopeful lad, "If he is not willing to sacrifice . . . everything, in order to devote himself to the development of his drawing ability then he had better not try."

Darling harbored no illusions about the power of the press in general or of cartoons in particular. In 1928, near the zenith of his fame, Ding told a gathering of life insurance company presidents: "A cartoonist may only play upon and reflect the things and emotions that are already before the public. He cannot successfully introduce new topics nor through his medium alone follow through with a process of reasoning. The cartoon is essentially a spotlight service. So it happens that he must be content to play his calcium ray upon the marionettes that strut the visible stage."

Two weeks following Pearl Harbor he wrote that the power of editorial pages and editorial cartoons had been greatly exaggerated, the state of the world being the proof. Ding believed cartoons were necessarily a primitive and very much abbreviated form of thought expression. He claimed they "are made for the day[,] and like flowers after picking they soon fade in interest." Many of Darling's cartoons have not faded but remain remarkably vivid. The parade that follows helps to demonstrate the impressive result of mixing Ding's artistic skill with his profound metaphorical sense.

The selections are generally arranged in chronological order. Some were chosen (from among the 15,000 he drew) because they are mentioned in the text of this biography. Others were selected because they illustrate recurring themes in Darling's work or represent views that he espoused with special fervor. Also included are examples of Darling's sketching, painting, and etching as well as a sprinkling of photographs and other illustrations that help to elaborate on the written record of Darling's life.

### The Rough Rider Takes a Few Shots at the Peerless Leader's Stock in Trade
John Henry (JH)

This signed drawing has sometimes been identified erroneously as Darling's first published political cartoon, but it was his third. Darling himself mistakenly recalled it as his first and improved it slightly for a book of his cartoons published in 1962 (right). Darling explained to the book's editor, John Henry, that he did not want readers to know his style had ever been so crude. The confusion between the first cartoon and this one was understandable since Teddy Roosevelt and William Jennings Bryan appeared in both. The cartoon originally appeared October 8, 1900.

**Sioux City Journal**

This photograph of the editorial room at the Sioux City *Journal* was taken near the turn of the century, when Jay Darling was a cub reporter, a photographer, and a budding cartoonist there. A. F. Allen, Ding's close friend, is shown at his desk.

## Now Look Pleasant, Please!

Drake University (DU)

Darling's first cartoon for his new employer, the Des Moines
*Register and Leader,* appeared December 9, 1906. The brash,
thirty-year-old cartoonist attracted protests from soft coal
dealers. The cartoon also excited representatives of the Roman
Catholic faith because Ding had mistakenly assumed Des
Moines was French for monk.

**Gracious Sakes, Theodore! If You Ain't Going In,
Please Get Off the Spring Board** (JH)

Darling focused on the conservationist President Theodore
Roosevelt again in 1908 when TR was questioning the wisdom
of his early renunciation of a second term. Roosevelt eventually withdrew from the race and anointed William Howard Taft
as the Republican standard-bearer.

### Just to Remind Those Who Are Not Aware of It that the Hunting Season Is Open
Iowa State Department of History (ISDH)

Darling's rapidly improving skill is obvious in this unsigned cartoon which appeared in the September 12, 1911, issue of the *Register and Leader*. As his first stay with the Des Moines paper drew to a close, few of his cartoons were signed. Those by Frank Moser, however, were regularly signed (right), even though they lacked the quality of Ding's work.

### Will He Ever Catch It?

**Well, So Long** (ISDH)

This cartoon appeared in the Des Moines *Capital* January 10,
1913, along with a brief front-page article reporting that the
ailing Ding was back on the job at the Globe syndicate after
surgery on his drawing arm. When W. B. Southwell, the
*Register and Leader*'s business manager, saw this he wondered
whether Ding would return to Des Moines. Ding did return to
the *Register and Leader* the following month.

(UI)

This portrait of Jay Darling
was taken about 1914,
shortly after his welcome
return to the Des Moines
*Register* following a brief
career at the New York
*Globe*. Darling never again
considered changing his
permanent residence and
thereafter considered Des
Moines and Iowa his home.

### The Annual Migration of Duck Is On (DU)

One of Darling's best-known conservation illustrations, this cartoon was his first under a ten-year contract he signed with the New York Herald Tribune syndicate. It appeared the day Darling signed the contract—October 21, 1916—on Darling's fortieth birthday.

BOYHOODS GREAT IDOL

## Gone to Join the Mysterious Caravan (DU)

Darling drew this cartoon to mark the death of William F.
(Buffalo Bill) Cody January 11, 1917. Cody, an Iowa native
who led hunts that helped to bring the American bison to the
brink of extinction, was an idol to many youngsters of the
nineteenth century. In retrospect, the western showman
represented the wholesale and wanton destruction of game,
which Darling later fought with vigor and determination.

## The Long, Long Trail (JH)

When his friend Teddy Roosevelt died January 6, 1919, Darling hurriedly sketched a farewell to the Rough Rider. Although Darling did not recall the source of the idea for the Roosevelt cartoon, its similarity to the earlier Buffalo Bill drawing is evident. Darling intended to do a more refined salute to TR for the Herald Tribune syndicate, but his hurried effort was greeted with such favor that a second was never attempted. This Ding Darling cartoon has been fashioned in more media and in more copies and is displayed in more public places than any other.

**The Operation Is About to Begin** (DU)

Darling's interest in medicine often came through in his cartoons. When the Iowa General Assembly convened for its 1917 session, Darling expressed the fears of many Iowa "patients" with this drawing in the January 8, 1917, issue of the Des Moines *Register*.

(DU)

(Opposite) When the first American casualties in World War I were reported (one of whom was Merle Hay of Glidden, Iowa), Darling drew this cartoon, which appeared November 8, 1917. On Armistice Day—November 11—in 1935 he revised the drawing (below, left). On the first anniversary of Pearl Harbor—December 7, 1942—he used the same theme to underline the human cost of war (below, right).

Bringing the Truth Home to Us

Lest We Forget in the Joys of Armistice
the Anguish Out of Which It Was Born.

(UI)

This photograph of Jay Darling was taken in his forty-second year (1918), approximately two years after his work was nationally syndicated by the New York *Herald Tribune*. Darling had become a famous figure and a significant asset to the Des Moines *Register*.

### There's a Man at the Door with a Package

(UI)

The United States geared up to provide food for the Allies during World War I, and with the end of hostilities there was an agricultural surplus. Darling showed the farmers' recurrent plight—low prices as a reward for efficient, high production—in this January 27, 1919, cartoon.

## High Time to Begin
(JH)

In 1948 Darling remarked that the only character "which has been completely of my own creation is the Iowa Farmer, old Uncle John Iowa. . . ." One of the earliest versions of the Iowa Farmer appeared in this cartoon June 26, 1919, when Ding urged Iowa to pull itself out of the mud. The more refined and stylized versions (below) appeared in later years. The model for Uncle John Iowa was Samuel H. Cook, a Van Meter, Iowa, implement dealer who died in 1932.

Des Moines *Register*

Darling was curious about airplanes and, characteristically, studied them and mastered their operation. Even after he no longer took the controls, Ding enjoyed traveling as a passenger in the several aircraft owned by the Register and Tribune Company. This photo was taken around 1920.

(JH)

DING"
'Y JAMES MONTGOMERY FLAGG

This sketch of Ding Darling was drawn about 1920 by James Montgomery Flagg, who was born the year after Darling and died two years earlier. Their careers were roughly parallel; they both enjoyed fame, artistic flair, and the company of famous persons. The flamboyant Flagg, however, preferred the company of Hollywood stars, while Darling enjoyed East Coast intellectuals. Flagg is best known for his sketches of celebrities and for his World War I recruiting poster, picturing Uncle Sam pointing to the viewer and saying, the "U.S. Army wants YOU!" Flagg, characteristically, described his Uncle Sam as "the most famous poster in the world."

(UI)

Jay Darling designed this spacious studio beside his Des Moines home. The unique structure, photographed in 1920, was situated on Darling's large and well-landscaped lot, attracted great interest, and was the subject of nationally circulated magazine feature articles. The pool at the right was one of the first private swimming pools in Iowa.

AN ORPHAN AT 8 IS NOW ONE OF THE WORLD'S GREATEST MINING ENGINEERS AND ECONOMISTS WHOSE AMBITION IS TO ELIMINATE THE CYCLE OF DEPRESSION AND UNEMPLOYMENT

THE SON OF A PLASTERER IS NOW THE WORLD'S GREATEST NEUROLOGIST AND HIS HOBBY IS GOOD HEALTH FOR POOR CHILDREN

A PRINTER'S APPRENTICE IS NOW CHIEF EXECUTIVE OF THE UNITED STATES

## In Good Old U.S.A. (JH)

Ding received his first Pulitzer Prize in 1924 for this cartoon which had appeared May 6 the previous year. Note that this drawing pays homage to Herbert Hoover as well as to Dr. Frederick Peterson and Warren G. Harding.

### Darn It! (UI)

Darling's humor pervaded most of his
cartoons in the 1920s. He generally drew
at least one each week that commented
on everyday occurrences in his readers'
lives. This is one example of Darling's
ability to put mass emotions into a small
space on a newspaper page. It appeared
January 19, 1924.

### Trying to Keep
### Pace with Events (DU)

In 1924 Darling com-
mented on the pace of
world events by showing a
harried historian astride a
winged steed in pursuit of a
rocketing earth. When,
after a year of illness, Darling returned to the drawing board April 5, 1926, he
used the theme again (op-
posite, left). The very next
day—April 6, 1926—he
noted, ''Things seem just
about where we left them a
year ago'' (opposite, right).
Note the dominance of the
prohibition issue in the latter cartoon.

This photo of Jay Darling was taken in 1925, when the cartoonist was approximately forty-nine years old. This was about the time Darling was stricken with peritonitis and was declared dead in obituaries published nationwide.

**If You Don't Think the World Moves Just Try Stopping a Year**

**Things Seem Just About Where We Left Them a Year Ago**

The Guild of FORMER PIPE ORGAN PUMPERS

*Pump, for the Wind is Fleeting*

KNOW *All Men by th*ese PRESENTS *that* Jay N. Darling, IS A QUALIFIED *and* ACCEPTED MEMBER WITH THE DEGREE *of* FELLOW PUMPER

*Dated at* NEW YORK JANUARY, 1, 1927

GRAND DIAPASON

Mary Darling Koss (MDK)

Jay Darling received dozens of awards and honors for his work as a cartoonist and conservationist. However, no award was more precious to him than this certification of his membership in The Guild of Former Pipe Organ Pumpers, presented to him in New York in 1927. Membership in the Guild was limited to persons who had actually pumped pipe organs by hand.

(MDK)

John Darling, Jay's son and alter ego, eventually became the medical doctor Jay had wanted to be. This photograph of John was taken in 1927 when he was a freshman at Princeton University.

## A Bumblebee's Nest

(UI)

This was one of Darling's most reproduced political cartoons because it told so succinctly what happened when President Calvin Coolidge twice vetoed the McNary-Haugen farm aid bill, which Congress passed twice and the farmers wanted. In spite of the backlash, Secretary of Commerce Herbert Hoover was elected to the presidency. This appeared April 26, 1928.

(UI)

**A Lot of Empty Chairs Were Left When Hoover Turned in His Resignation As Secretary of Commerce**

(DU)

The owners and editors of the New York *Herald Tribune* visited Darling in Des Moines in 1928 and took time to pose with the syndicate's star cartoonist. Geoffrey Parsons is at far left. Next to him are Ogden and Helen Reid. Darling is at far right. The man next to Darling is Arthur Draper. Whitelaw Reid, son of Ogden and Helen, recalled that Ding was a by-word for excellence in cartooning. "Both my mother and father," he wrote, "thought of him as being at the very top of his profession. All other cartooning seemed to be either a pale imitation of his work or of a totally different school. . . ."

Darling and Herbert Hoover were close friends for many years, and Ding's regard for his fellow Iowan was obvious in nearly all his Hoover cartoons. When Hoover resigned as Secretary of Commerce in 1928, Darling marked the occasion with a cartoon. When Hoover was inaugurated as President in 1929, Darling again saluted his friend.

**Only in U.S.A. Does This Happen**

## Waiting for Saturday Night

(JH)

A June 1, 1931, cartoon expressed Darling's disenchantment with "the noble experiment" that had brought crime and corruption in its company. Note the error on the calendar—1913 should have been 1931. A March 15, 1929, cartoon scolded social-drinking scofflaws for their contribution to organized crime.

## Such Familiarity

**The Smallest Species of Deer in North America, Alone,
Unguarded and on the Way Out**          Darling Papers (DP)

This evocative cartoon was drawn by Darling on a lap board as
he sat in a boat near one of the Florida keys. Drawn about
1930, it was helpful in gaining passage of a Key Deer Refuge
Bill in 1957.

This photo of Jay Darling, standing near corn stubble wit
double-barreled shotgun, was taken in South Dakota in 19
The energetic outdoorsman was about fifty-five years

(MDK)

Mary Darling, Jay's daughter, was born in New York during Ding's otherwise disappointing and brief career with the New York *Globe*. This photograph was taken in 1930, when Mary was a student at the Emma Willard School.

South Dakota Hunt,
with J.S. Carpenter 31

**Surely These Two Overburdened Men Ought to Be Able to Understand Each Other** (UI)

This sympathetic cartoon was published October 4, 1932— exactly one month before President Herbert Hoover was swept from office by Franklin Delano Roosevelt's landslide election victory.

### Why Not a Nobel Prize for Scientific Discovery to the Hen? (UI)

Shortly after Roosevelt's victory, Darling expressed his distrust of the New Deal with this comment on farm production limitations suggested by Henry A. Wallace, FDR's secretary of agriculture.

### Some Folks Won't Follow Leadership When They Have It (UI)

This cartoon appeared April 14, 1933, shortly after the Roosevelt administration had taken office. Henry A. Wallace was a neighbor and friend of Ding's and, in spite of some political differences, Darling gave Wallace support in the early years of the New Deal.

**We Must Learn to Employ Our Idle Time with Other Things Than Work**
**or**
**Introducing the New Franklin D. Roosevelt Theory of Economics and Education**

(MD

Ding's fans had their favorites among the thousands of cartoons he drew in his 49-year career. This cartoon, published i
*Collier's* magazine in 1934, was a favorite of Darling's.

(UI)

Jay Darling was appointed
chief of the Biological Survey
March 10, 1934, by Secretary
of Agriculture Henry A.
Wallace. When Darling
(center) was sworn into office,
Wallace (left) was at his side.

Darling's conservationist instincts were obvious in this March 13, 1934, cartoon. The drawing was published just three days after Darling had been officially appointed chief of the Biological Survey.

Shortly after he became chief of the Biological Survey in 1934, Jay Darling (center) purchased the first duck stamp, which he had designed. Washington, D.C., Postmaster William Mooney is at left, and C. B. Ellenberger is at right.

(DP)

One highlight of Darling's brief career as chief of the Biological Survey was this illustrated appeal in 1935 to President Franklin Roosevelt and Roosevelt's pointed reply (turn page) to the effect that Darling had raided the treasury.

HEY! LOOK OUT WHAT YOU'RE DOIN'!

JAY N. DARLING
CHIEF, BUREAU OF BIOLOGICAL SURVEY

MY $6,000,000

(OR SO REX TELLS ME)

My dear Mr. President:

We can make better use of retired agricultural land than anybody.

Others just grow grass and trees on it. We grow grass, trees, marshes, lakes, ducks, geese, furbearers, impounded water and recreation.

The six million we got from Congress  and

which you think is enough, is mostly going
to buy Okefenokee, the ranches on the winter elk range in Jacksons Hole, the private
lands that lie in the midst of the Hart Mt.
antelope range, and for rehabilitation
(dams and dikes) of the duck ranges we
bought last year.

By the way, Secretary Ickes wants me to
give him Okefenokee. Do you mind? I don't,
only that it cuts into our nesting area
funds.

I need $4,000,000 for duck lands this
year and the same bill which gave us the
$6,000,000 specifically stated that at your
discretion you could allocate from the
$4,800,000,000 money for migratory waterfowl
restoration.

We did a good job last year. Why cut
us off now?

July 26, 1935.

THE WHITE HOUSE
WASHINGTON

July 29, 1935.

Dear Jay:-

     As I was saying to the Acting Director
of the Budget the other day - "this fellow Darling
is the only man in history who got an appropriation
through Congress, passed the Budget and signed by
the President without anybody realizing that the
Treasury had been raided."

     You hold an all-time record.  In addition
to the six million dollars ($6,000,000) you got,
the Federal Courts say that the United States
Government has a perfect constitutional right to
condemn millions of acres for the welfare, health
and happiness of ducks, geese, sandpipers, owls
and wrens, but has no constitutional right to
condemn a few old tenaments in the slums for the
health and happiness of the little boys and girls
who will be our citizens of the next generation!

     Nevertheless, more power to your arm!
Go ahead with the six million dollars ($6,000,000)
and talk with me about a month hence in regard
to additional lands, if I have any more money left.

                 As ever yours,

                 Franklin D. Roosevelt

Honorable J. N. Darling,
Bureau of Biological Survey,
South Building,
Washington, D. C.

### Did Someone Say He Wasn't Well? (JH)

Darling was not a political friend of Franklin Roosevelt's, but he penned this pro-Roosevelt cartoon, which appeared August 30, 1935, just a few months before resigning as chief of the Biological Survey. The figure behind FDR is Marvin McIntyre, the president's secretary and a friend of Ding's.

(UI)

When Jay Darling resigned as chief of the Biological Survey in November 1935, his colleagues and friends in the department presented him with a new shotgun. Ding is shown with his successor, Ira (Gabe) Gabrielson.

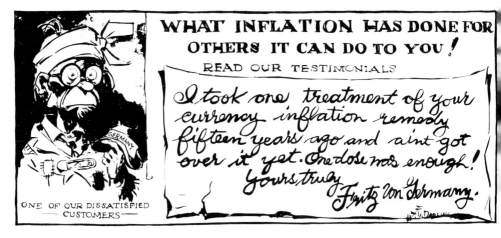

**Unfortunately Inflation Is a Patent Medicine** (MDK)

Another Darling favorite was this cartoon, also done for *Collier's* magazine in the mid-1930s.

**Never Mind the Political Souvenirs. Get a Doctor** (JH)

The Supreme Court ruled Henry A. Wallace's Agricultural Adjustment Act unconstitutional January 6, 1936. Darling responded with another cartoon favorable to the secretary of agriculture under whom he had worked as chief of the Biological Survey.

## Disturbing to the Occupant
(JH)

This comment, published on Halloween in 1936, accused President Franklin Roosevelt, Harry Hopkins, and Postmaster General James Farley of usurping the private rights of U.S. citizens. Although many subscribers to the Herald Tribune syndicate service published the drawing, it did not run in the *Register* because publisher Gardner Cowles considered the cartoon indelicate and in poor taste.

## Trailers Are Swell—The Trouble Is to Find a Place to Put 'Em
(DU)

In one of the many illustrations for Darling's *The Cruise of the Bouncing Betsy* he lamented the lack of adequate trailer parks.

## Why Call Them Sportsmen?

(DU)

Another of Darling's most popular cartoons was this one in defense of defenseless ducks. As chief of the Biological Survey, Darling was tough on game law violators and earned a reputation as the best friend a duck ever had.

## The Kidnaping—Or More and More Democracy

(JH)

When Secretary of the Interior Ickes moved to put the Forestry Service and the Biological Survey within his department, Darling pictured his friends— Ira Gabrielson, Gus Silcox, and Henry A. Wallace—being held silently at bay by FDR. The cartoon was published November 21, 1937.

**How Rich Will We Be When We Have Converted All Our Forests, All Our Soil, All Our Water Resources, and All Our Minerals into Cash?** (UI)

This cartoon has been used often by conservationists because it illustrates so well the need to save natural resources—the source of all wealth. It appeared June 26, 1938.

(UI)

Darling is shown at his drawing board in this 1938 photograph. Darling did much of his reading, writing, and thinking at his drawing board and preferred the board to a desk.

### The Dubious Rewards of Neutrality (JH)

This pointed Darling comment on international neutrality appeared six months before Darling and several other prominent persons appealed to Franklin Roosevelt to enter World War II in support of the Allies. The cartoon was published January 25, 1940.

### The Forty-Hour Week (JH)

Even though the United States was not directly involved in World War II, the nation's factories were supplying war materials to the Allies. Darling showed his disdain in 1940 for labor and its spokesmen's demands for shorter hours by comparing the forty-hour week with twenty-four-hour-a-day warfare in Europe.

(MDK)

Darling's Fish House (below) was constructed off Captiva Island in 1941. Set in a paradise, it was designed by Darling and for years afforded him the occasional solitude and privacy he wanted. It also was a commodious place where he and Penny could entertain their many friends and members of their family.

**What a Place for a Waste Paper Salvage Campaign** (JH)

When Darling won his second Pulitzer Prize for this cartoon, which appeared September 14, 1942, he had trouble recollecting it and questioned the judgment of the panel members who had chosen it as the best U.S. cartoon of the year.

(MDK)

This undated photograph of Jay Darling at his drawing board was probably taken in 1942.

**An Urgent Plea** (UI)

As World War II drew to a close, Darling fashioned this illustration May 9, 1945, and included it in a proposal for the creation of a conservation clearing house. The proposal began with this illustration, accompanied by the words, ''200 years of exploitation climaxed by two world wars have left quite a hole.''

It's Hard to Start a Fire with One Stick of Wood

But if You Could Ever Get the Fire Wood Together in One Pile

(UI)

This illustration, created May 31, 1945, was not used in the *Register* or by the Herald Tribune syndicate but in the conservation clearing house proposal, which urged an investment of $100,000 to set up an organization for three to five years, after which time it was expected to support itself.

### Eventually, Why Not Now? (JH)

This cartoon, which appeared November 9, 1945— approximately two months following the end of World War II— underscored Darling's continued support of an international body to enforce world peace. The need was even more pressing than when the United States refused to become a member of the League of Nations, Darling intimated, because World War II had introduced the specter of atomic warfare. Darling told his friend, John Henry, that this cartoon was Pulitzer Prize material and was disappointed when it was not selected for the award. Frank Miller, a Darling successor at the *Register*, won a Pulitzer Prize in 1963 for a January 15, 1962, cartoon expressing a similar theme (below).

### I Said—We Sure Settled that Dispute, Didn't We?

**Add Crime Wave** (JH)

With this cartoon, published in the *Register* August 21, 1946, Darling lambasted Iowa's Governor Robert Blue and his state Republican chairman for what Darling insisted was their political interference in the operation of the Iowa State Conservation Commission, which Darling had helped establish as a nonpolitical agency.

### The Only Kettle She's Got (UI)

This August 6, 1947, cartoon was probably Darling's most effective description of the effects of an expanding world population on world food supplies.

**Thank Goodness Now We'll Know Where They All Are** (DU)

When Henry A. Wallace—Harry Truman's former vice-president—split with the Democratic party and ran in 1948 for President on the Progressive party ticket, Darling reflected the sentiments of many readers with this January 13, 1948, cartoon.

### It's Nice Somebody Can Enjoy It (JH)

Darling foresaw shortages of fossil fuel long before they became generally apparent. This 1948 cartoon, drawn twenty-five years before the 1973 oil embargo, has currency in the late twentieth century, which is the case with most of Ding's cartoons that scold Americans for their waste of natural resources. Note the dedication to John Henry in the lower left-hand corner.

### Sing Brothers, Sing! (JH)

Darling occasionally dedicated a cartoon to an individual or an organization, with no intention of having it published. Just as he dedicated the "gas shortage" cartoon to John Henry, about 1949 he gave this one to the Press Columnists of Iowa at Henry's suggestion. Members made it a traveling trophy awarded each year to the organization's president.

**Tortugas Day Shift Shrimpers**                          (UI)

Darling, honored for his cartooning artistry, was skilled in
other media of artistic expression. He was an accomplished
whittler, he dabbled in water colors, and he created exquisite
etchings. This water color was executed in 1951 off the Dry
Tortugas, just two years following Ding's retirement as the
*Register*'s cartoonist.

Jay and Penny Darling were photographed at the rail of the
S.S. *Constitution* with Vernon Clark—Jay's co-worker in his
earliest days at the Des Moines *Register*—as they departed for
a Mediterranean tour in 1952. Genevieve Pendleton Darling
was a petite woman who modestly embodied a large intellect.
She graduated from the University of Wisconsin in three years,
studied at the University of Chicago, and worked as a physical
therapist in Boston before marrying Jay Darling in 1906. She
was independent and strong willed, loved to travel, and after
her marriage often did so alone. Penny worked energetically to
keep Jay's home and family life in order. She kept "busy as a
little red wagon," as Darling often put it, and her self-reliance
gave Ding great freedom to pursue his dual career as cartoonist
and conservationist.

### Cartoonist Decorates a Building (JH)

Nine Darling drawings were etched in metal and permanently installed above the main doors of the University of Iowa Library in 1953. The cartoon series took a lighthearted view of the history of knowledge and its results. When the library was later remodeled, the cartoons were moved inside to a fourth-floor hallway.

**1** A hirsute cave man started it all.

**2** Then the Egyptians wrote in stone.

**3** This was how 'twas done in Rome.

**4** Prolific was the copyist monk.

**5** Then, at last, came the inky press.

**6** One result: the lawyer's tome.

**7** Another: the treatise scientific.

**8** The library amasses all this lore.

**9** It's squirted into student's dome.

**Can't You Find Any Other Place for Your Target Practice?**

(DU)

Although Ding steadfastly refused to do any cartooning following his retirement from the *Register* in 1949, he relented in 1955 when the Department of Defense, headed by Charles Wilson, threatened to make an artillery range of the Wichita Mountains Wildlife Refuge.

Darling exercised skill and patience in the execution of his etchings. These are pencil studies done in preparation for the painstaking etching process.

Iowa State University (ISU)

The largest single collection of Darling etchings—a total of sixty-two—has been permanently displayed since 1975 in the Scheman Continuing Education Building at Iowa State University. The Darling Lounge, adjacent to the Brunnier Gallery, is available to many thousands of visitors each year.

*Explosion in the Lily Pads*

For Ding the perfectionist, etching was no more than a pleasant pastime and the resulting prints were products of his hobby. But Darling's great artistry—which brought vitality, movement and detail to so much of his work—is especially evident in his beautiful etchings.

*After Closing Hour They All Come In*

*Design for the First Federal Duck Stamp*

The original of this etching was given to Dr. Norman E.
Borlaug, food scientist and Nobel laureate, as a symbol of the
Iowa Award presented in February, 1978. The etching was
donated by the J. N. ''Ding'' Darling Foundation, Inc.

### 'Bye Now—It's Been Wonderful Knowing You  (JH)

Darling's famous farewell cartoon appeared in the Des Moines *Register* February 13, 1962, the day following his death. Darling had drawn the cartoon in 1959 during a serious illness. Among the many other symbols in the drawing, note Ding's famous Teddy Roosevelt, the Iowa Farmer, and the duck stamp design on the wall as well as the flying goose wildlife refuge symbol on the cabinet door. Note also Ding's fishing rods, gun case, and decoys gathering cobwebs behind the couch.

The Conservation Center—a project of the privately supported Sanibel-Captiva Conservation Foundation—provides exhibits, educational programs, and more than four miles of wetland nature trails that help enhance understanding of the need to preserve and protect the remaining unspoiled areas of the islands. The accomplishments of Ding Darling, one of the islands' most famous inhabitants and one of the nation's best known conservationists, are prominently recognized at the center, which is located near the J. N. "Ding" Darling National Wildlife Refuge formally dedicated in 1978.

*The Island Reporter*

The National Wildlife Refuge on Sanibel Island, named in Ding Darling's memory in 1967, was officially dedicated in ceremonies conducted on the island February 4, 1978. A plaque on the Darling National Wildlife Refuge headquarters (below) reads in part, "It was Darling's hope that future generations could share the beauty, serenity, and the bounty of nature he had known."

J. N. "DING" DARLING NATIONAL WILDLIFE REFUGE

ESTABLISHED IN 1945

J. N. "DING" DARLING WAS A RENOWNED CARTOONIST AND ECOLOGIST WHOSE PERCEPTIVE MIND, ELOQUENT PEN, AND SKILLED BRUSH ENDEARED HIM TO NEWSPAPER READERS AND CONSERVATIONISTS DURING A LONG CAREER, AND A LIFETIME OF ECOLOGICAL CONCERN.

"DING'S" SYNDICATED CARTOONS, WHICH BROUGHT HIM INTERNATIONAL FAME AND TWO PULITZER PRIZES, OFTEN DEALT WITH WATERFOWL AND THE CONSERVATION OF NATURAL RESOURCES.

HE WAS ONE OF THE EARLIEST SUPPORTERS OF THE MIGRATORY BIRD HUNTING STAMP PROGRAM. AS CHIEF OF THE BUREAU OF BIOLOGICAL SURVEY, FORERUNNER OF THE U.S. FISH & WILDLIFE SERVICE, HE DESIGNED THE FIRST FEDERAL "DUCK STAMP" AND THE FAMILIAR NATIONAL WILDLIFE REFUGE FLYING CANADA GOOSE SIGN.

FOR MANY YEARS DARLING HAD A WINTER HOME ON CAPTIVA ISLAND. HE INITIATED AND, TO A GREAT EXTENT, WAS RESPONSIBLE FOR THE ESTABLISHMENT OF THE SANIBEL NATIONAL WILDLIFE REFUGE, RENAMED IN HIS MEMORY IN 1967.

IT WAS DARLING'S HOPE THAT FUTURE GENERATIONS COULD SHARE THE BEAUTY, SERENITY, AND THE BOUNTY OF NATURE HE HAD KNOWN.

# 10

## Russia, Relatives, and Resources

JAY DARLING was not eager to run for the U.S. Senate, but some of his supporters may have been falsely encouraged when, at about the same time the "Ding for Senator" movement was in full swing, he left for the Soviet Union. The famous American cartoonist had been invited to visit the U.S.S.R. by Joseph Stalin and gladly took advantage of the opportunity. He traveled around the country for five weeks accompanied only by a non-Communist interpreter. Of course, he kept his sketchbook at the ready and carefully recorded his observations.

Merle Houts's first assignment when she began working for Darling was typing the manuscript for *Ding Goes to Russia,* published in 1932. Houts had been told when she applied to the *Register*'s employment office that there was some work to be done, that it would require great care, and there was no knowing whether it would last very long. Darling's trip was an eye-opening experience. "That my Russian trip should have resulted in a book is a complete surprise to me," he wrote, "but if it serves to quiet some of the hysterical fears of others, as it did mine, it will have served a good purpose."[1]

The *Register* cartoonist was surprised at the absence of militarism and violence in the Soviet Union. "There is about as much Communism left in Russia as there is tobacco in their cigarets," Darling quipped. "Most of the Russian tobacco, like their Communism, seems to be raised for export purpose," he suggested. The author advised every American to make a trip to Russia at least once, "just to see what happens when the upper crust of society gets too heavy and overbearing and the proletariat rises up and gives it the axe." Less than two decades after the revolution Darling discerned a failure of the new ruling class to make its ideas work. It was beneficial, he wrote, to learn how many holes were poked in the favorite theories of the utopian reformers when those theories were put to the test. Darling noted that the new banker looked like the man who used to take care of the furnace. The bank cashier, he said, might easily still be mistaken for the scrubwoman, and some of the clerks in charge of foreign exchange must have been night watchmen before the revolution. ''I have nothing against night watchmen, scrubwomen or furnace men,'' he observed, ''but they certainly make funny bankers when they begin.''[2]

Darling's Russian trip had a lighter side. When he was in New York, Ding purchased three pairs of pajamas in preparation for his departure. Once within Russia's borders he caused a minor stir, after which it was brought to his attention that the pattern on his new pajamas was similar to the executed Czar's coat of arms. He thereafter was careful to conceal his inflammatory sleepwear.

Because Jay was busy with his Russia project and Penny was in Europe, neither was on hand for Mary's graduation from Emma Willard School. John, meanwhile, had graduated as a Phi Beta Kappa and cum laude from Princeton. Even before graduation exercises John departed for Rush Medical Center in Chicago, where he at once began to pursue the medical education Jay himself had once wanted. The brilliant young Darling was on his way to becoming an exceptional doctor.

Mary went to Scotland after taking her college boards and later met her mother in Switzerland. Jay, who had been in Russia, joined them in Hamburg and the trio went to Norway, Denmark, and Sweden. Darling thoroughly enjoyed traveling Europe in comfort. He occasionally mentioned the contrast between the confidence of affluence and the embarrassment he felt when his father could not afford tips, leaving the Darling family to carry most of its own luggage on Ding's first European tour. Mary returned from Europe to enroll at Vassar College, where she studied for two years before marrying Richard B. Koss of Des Moines.

While Jay Darling and his family were in agitated motion, so was Jay's older brother Frank. Slightly taller and heavier than Jay, Frank greatly resembled the famous cartoonist. On his visits to downtown Des Moines, on his way to or from the Register and Tribune Building, Frank became accustomed to returning greetings intended for his brother.

Frank had proved his intellectual mettle in school and his ability to get things done in education and business. His erratic career was as scintillating in its own way as Jay's had been. Frank had graduated from high school in Sioux City, as had Jay. He then attended Grinnell College, his first step toward the medical degree he too had set his sights upon. He graduated in 1895 from Grinnell with Phi Beta Kappa honors. While an undergraduate, he had played football and had been manager of the baseball team and editor of the college annual. He surveyed and established the first golf course in Sioux City and taught science and biology in high schools in Aurora and Oak Park, Illinois.[3] He later joined the Chicago school system and progressed steadily from superintendent of physiography in the city's schools to superintendent of Chicago's vacation schools. He taught biology at Chicago University Teachers College and studied medicine for two years at Rush Medical Center, where his nephew John later enrolled.

While in the midst of his medical school career, Frank fell in love with a charming woman, and it marked the end of his hopes for a medical degree. At her insistence he gave up medical school to marry her, and "he had a pretty tough time finding any other field of occupation adeqate to their needs."[4] But not for long. Frank Darling became a manager for the Clay Products Company in Terre Haute, Indiana. Soon afterward, Lamarcus Thompson, an old family friend, took a hand in his life and career. Thompson, with whom the Reverend and Mrs. Marc Darling had remained friends following their ministerial tour of duty in Elkhart, Indiana, was a remarkably inventive and resourceful entrepreneur. Thompson had invented the roller coaster. He created the undulating marvel following a trip west, where he had watched as mining buckets came bouncing down a hillside cable. It occurred to him that a ride that provided similar motions would be a popular amusement. His L. A. Thompson Scenic Railway Company became the umbrella organization for holdings in Rockaway Beach, Coney Island, Philadelphia, Long Beach, and San Diego amusement parks.[5] Thompson also invented seamless stockings and sold his first order to Marshall Field, founder of the famous store that bears his name.

Thompson, whose comic postcard had helped launch Jay Darling's cartooning career, hired Frank to oversee the installation of roller coasters throughout the world. Following Thompson's death, Darling became president of the vast Thompson Scenic Railway Company and made a fortune looking after its interests in the United States as well as in London and Paris.[6]

For four years, Frank was president of the National Amusement Association. He resigned that position to oversee the building and operation of the Playland amusement area at Rye Beach, New York, which was a very profitable venture. He later superintended the building and inauguration of amusement and guest accommodations for the famous Rainbow Room at Rockefeller Center, New York. He so impressed the Rockefeller family with his talents that he was put in charge of guide services and entertainment in conjunction with the reconstruction of Williamsburg, Virginia. He also administered the children's amusement parks at the world's fairs in Chicago (1933–1934) and New York (1939–1940). Wrote Jay, "He was quite a genius in the field of organized devices for the amusement of the public in parks and public playgrounds."[7]

Frank, who was jovial, inventive, and energetic, was a poor manager of his resources. And, according to Jay, Frank was an equally poor judge of the women he chose to marry. Possibly, Jay revealed, Frank's devotion to his consuming interests left his first wife with too much time for other men. Following a period of "supreme unhappiness," the couple was divorced. "The second wife," Jay recounted, "captured him in the period of emotional distress which followed, and except for his contact with old friends and when he was away from home I doubt if Frank ever really got much pleasure out of life." Frank's considerable wealth, between the two wives, was stripped from him. But "thank fortune," wrote Darling, "I had been sufficiently successful so that Frank never was in need." Jay fondly summarized the life and character of his older brother in a letter written two years after Frank's death:

Frank's over-generous and charming personality did not profit him greatly in the later years of his life but when you knew him [as a young man] and as I like to remember him, he was a great gift to happy companionship. . . . His going was a "happy release" from a very unhappy situation. That situation might possibly be most accurately described as a very painfully critical and gold-digger second wife, who had made the last twelve or fifteen years of his life a perpetual torment.[8]

Nearly a quarter-century following Frank's death, the fiftieth anniversary of the opening of the famous Coney Island Cyclone penny-a-second roller coaster was observed with several hours of free rides. It was reported that the Cyclone had entertained ten million riders in its fifty years and that it was a favorite of aviator Charles Lindbergh. It is likely that none of the celebrants had ever heard of Lamarcus Thompson or Frank Darling.[9]

Although Frank and Jay shared the congenital Darling drive, intelligence, and affability, their careers took them in seemingly opposite directions. While Frank busied himself making several fortunes in the business of mass amusement and escapism, Jay gave increasingly of his time to the crusade to save the nation's natural resources and to halt the indulgence and frivolity that led to unnecessary waste. Ding enjoyed life and was blessed with a rich sense of humor, a trait that showed repeatedly in his cartoons and his writing. His joy vanished, however, when he saw irreplaceable or limited resources squandered.

When Darling returned from most of the summer of 1931 "on my hands and knees going over Russia," he attended his first meeting of the Iowa State Fish and Game Commission. Dr. W. C. Boone of Ottumwa was presiding as chairman of the commission. The state's projects were under discussion and the meeting was moving along well.[10]

Iowa Governor Dan Turner, whom Darling had upstaged at the Republican convention, recalled with pride thirty years later that he had appointed Darling chairman of the commission, but he could not recall whether Darling had been the commission's first chairman. At his first commission meeting Darling guessed that the other members had previously elected Dr. Boone to the post, even though some of the members assumed that Darling had been named chairman by the governor. Darling commented, "I had made some enemies in the conservation field and I didn't want to stir up the muddy waters and rather than make any contention of Boone's chairmanship I just went along with the program and there was no expression of dissatisfaction that I heard." He credited Turner with having "started me on my way" and wrote that, without Turner's backing of important Iowa conservation policies, the story of his own role in national conservation efforts "would never have been written."[11]

In 1932 Iowa's Fish and Game Commission came face to face with the fact that there were not enough scientifically trained persons to do a professional, nonpolitical job of wildlife research and administration. In typical, direct Darling fashion the commissioner-cartoonist paid a visit to R. M. Hughes, president of Iowa State College (now Iowa State University) and proposed that the college join in a cooperative program for research in wildlife conservation. The Darling proposition called for tripartite support. The college was asked to provide $3,000 a year for three years. Darling said that he would secure an additional $6,000 a year from other sources to put the revolutionary idea into practice. He must have been persuasive because Dr. Paul L. Errington, a Ph.D. graduate from the University of Wisconsin and a student of the venerated Aldo Leopold, joined the Iowa State College staff July 1 of that same year as leader and organizer of the nation's first Cooperative Wildlife Research Unit.

Darling had swung the deal by negotiating a three-year agreement among the college, the Fish and Game Commission, and himself. Eager to put his money where his convictions were, Darling personally pledged $3,000 annually for the term of the contract—a total of $9,000 in depression-weary 1932. The college provided facilities, services, and salaries evaluated at about $10,000 per year. That, combined with the contributions from Darling and the commission, got the research unit off to its unpretentious beginnings. During its first few years the office and research space provided for the unit leader consisted of a room in the Insectary Building. The room housed Dr. Errington, his graduate students, bags of owl pellets, a reference collection, and piles of rubber boots. The success of the pioneering Cooperative Wildlife Research Unit at Iowa State College later provided Darling with a springboard for instituting the scheme on a national scale.[12]

I. T. Bode, a graduate of Iowa State College in forestry, was serving as chief game warden for the state at the time the Iowa research unit was established in 1932. In 1935 the Fish and Game Commission and the state Park Board were merged in a new seven-member state Conservation Commission, and Bode was succeeded by Fred Schwob, an enthusiastic supporter of the research unit.[13]

One of the first projects of the Iowa Fish and Game Commission was also one of the most farsighted undertaken in any state. Under the leadership of Professor Leopold of Wisconsin, work was begun by some of the best brains in the field on a comprehensive twenty-five-year conservation plan for Iowa. The plan was one of the first of its kind in the nation. The merging of the Fish and Game Commission and the Park Board grew out of recommendations made in the twenty-five-year plan.[14]

Darling noted years later that an appropriation of $35,000, to be derived from license fees, had been made to fund the complete biological survey of the State of Iowa, its rivers, its marshes, its water resources, and its upland game prospects; and the long-term program was set up on the basis of that survey. Darling lauded those involved: "It was the first complete job of its kind in the United States and but for the untiring efforts of the Izaak Walton League and the peculiar susceptibilities of Governor Dan Turner to their evangelism this pioneer job of surveying and planning for a definite program of conservation could not have been accomplished."[15]

Through his incisive cartoons and vigorous successes in Iowa, Darling had established himself as a conservationist of the first order—a fighter, a mover, an articulate force. He possessed a gift for posing conservation dilemmas in the most commonsense terms. He had for years simplified complex national and international issues, including pollution and stewardship, by proceeding from the known to the unknown— by allowing his reader to view unfamiliar territory from familiar turf.

Meanwhile, a series of drought years had played havoc with water holes, lakes, and streams in the nation's midsection and had brought with them desperate conditions for wildlife in the area. Franklin Roosevelt, an avowed conservationist, was president following a tidal wave vote that had washed Darling's friend Herbert Hoover out of the nation's highest office. Members of Roosevelt's cabinet openly boasted

that the Franklin Roosevelt administration would outshine the spectacular conservation achievements of the Teddy Roosevelt administration.[16] But FDR needed help on the wildlife conservation front.

Roosevelt's secretary of agriculture was Henry A. Wallace, a friend
of Darling's and like Ding a resident of Des Moines. Darling had been
involved in party politics with Wallace when the secretary had been a
Republican and took at least part of the credit for turning him into a
Democrat. The incident had grown out of the Republican convention of
1932, where Darling had served on the Resolutions Committee. "That
was the year when Prohibition and the Depression were the absorbing
topics," Darling later wrote, "but there was a brief and meaningless
plank on the agricultural subject." Darling objected to this, and the
chairman of the committee appointed a subcommittee of five to rewrite
the plank. At lunchtime, the subcommittee retired to do its work.
When its members returned, "completely sold" on the two or three
issues Darling thought most important, the Resolutions Committee was
embroiled in a heated discussion of Prohibition, which went on until
about 4:00 P.M. Finally the chairman congratulated the committee
members on their work and, noting there was nothing more to add to
the platform, said he would welcome a motion to adjourn.

Darling was outraged: "With the special agricultural committee
farm plank in my hand I arose and protested that the report of the
special committee on agriculture had not been called for and asked for
the privilege of presenting the farm plank." The surprised chairman
replied that the farm policy had already been adopted. He had found a
copy of the amended plank on his table following lunch. It had been
approved and sent to the printers. Ding and his subcommittee had been
sabotaged. Darling went to the phone to report to Wallace, his mentor
in agricultural matters:

At that time Henry Wallace was a Republican and I had called him on long-dis-
tance and read him the amended farm plank as the special committee had written it. He thought our farm plank was all right and as I look back upon it I
presume many of the ideas expressed [in it] had come out of the numerous ses-
sions I had had with Henry before going to the convention. When Henry heard
my story of what happened in the Resolutions Committee he promptly
repudiated his membership in the Republican ranks and turned Democrat,
campaigned vigorously and became Secretary of Agriculture.[17]

While Darling respected Wallace and was willing to give the future
vice-president of the United States the benefit of the doubt, his low
regard for FDR was a secret to no one. Ding's cartoons, his conversa-
tions, and his letters made his distrust of Roosevelt's liberal ideas un-

mistakable. He saw in Roosevelt's policies the destruction of private initiative and enterprise and the beginnings of a welfare state.

Most observers would have considered outrageous and fantastic any suggestion that Darling would help execute one of the most significant aspects of Roosevelt's conservation program. And yet that is exactly what was about to happen.

# 11

## *Cartoons to Conservation*

JAY DARLING in later years freely admitted that his friend Herbert Hoover had done nothing about the threat to wildlife caused by years of drought and that his enemy Franklin Roosevelt did recognize the problem. Part of Roosevelt's response was to appoint a special presidential committee of three men to study the situation and to recommend a program for restoration, or conservation, of migratory waterfowl. The men appointed to the committee were Tom Beck, editor of *Collier's* magazine and chairman of the Connecticut State Board of Fisheries and Game; Dr. John C. Merriam of the Smithsonian Institution in Washington, D.C.; and J. N. Darling. The committee was appointed in January 1934 "to devise a wildlife program that would dovetail with his [the president's] submarginal land elimination program."[1]

Merriam was unable to serve on the committee and was succeeded by Aldo Leopold. Darling was appointed because of his long interest in conservation, his leadership as a member of the Iowa Fish and Game Commission, and his experience as a member of the advisory board that through the years had made recommendations for waterfowl seasons and bag limits to the federal authorities. Beck, the chairman of the committee, had proposed to Roosevelt in October 1933 that an allotment of $12 million from employment relief funds be made for a new wildlife restoration program. Beck had also proposed that the restoration program, so funded, be run by an administrator independent of existing government agencies.[2]

The president had asked Secretary of Agriculture Henry A. Wallace to evaluate the proposals, and Wallace in turn forwarded them to Paul G. Redington, then chief of the Biological Survey. The Bureau of Biological Survey, forerunner of the U.S. Fish and Wildlife Service, in 1934 was an arm of the U.S. Department of Agriculture under

Wallace's direction. Chief Redington had directed a national survey of waterfowl in the late 1920s, which showed that prompt action was needed to preserve them for future generations. As one result the bag limit was cut from twenty-five to fifteen ducks, effective December 31, 1929; but the waterfowl population continued to plummet.[3]

Redington's long memorandum of November 10, 1933, informed Wallace that the majority of Beck's suggestions had been studied, that the Survey had made plans to see the projects accomplished, and that what was needed most was the financial support authorized in the Federal Wildlife Refuge Act approved by Congress in February 1929. That act authorized a total of $8 million over a ten-year period. Because of the Depression, however, the funds were never appropriated. Rex Tugwell, deputy secretary of agriculture, in his letter of transmittal to the White House, commented that "there is nothing novel in Mr. Beck's proposal." He suggested that the president might seriously consider the allocation of anywhere between $10 million and $15 million for the acquisition of further game refuges and pointed out that Beck's suggestions made no mention of regulating the annual kill of waterfowl.

Once appointed, the so-called Beck committee worked feverishly at its assignment, as Darling later recalled:

Out of that Committee came the first enunciation of the essential program of restoring ducks' nesting grounds, it being recognized that the drainage program in the duck nesting areas had reduced the nesting areas to a very small percent of their former acreages. On the basis of that conclusion, we solicited advice from every state conservation officer or commission in the whole United States as to areas which might be subject to restoration, and laid out quite a program of possibilities. It was a hasty job and we knew at the time that much of the ground which we recommended for restoration was in agricultural production and probably too high priced for restoration to marshes and lakes. We sought, of course, the advice of the Biological Survey and asked them to furnish a program, they being the one authoritative source which should have had material ready to submit.

Darling characterized the report provided by the Biological Survey as "a very poor thing" and suggested that its quality may have been influenced by the prominence accorded the Beck committee. It may also have resulted from the Survey having no program of its own. "I was never able to decide," Darling reported. He was disappointed that the Survey was able or willing to provide little of value. "In fact," wrote Darling, "their areas designated as nesting ground refuges were less qualified than most of those that we got from the various state conservation officers."

We got very little information out of the members of the Bureau and because they refused to deliver any plans for an extension of the Refuge system or restoration of the nesting grounds, we had to ask Henry Wallace to requisition, for his inspection, any plans which they might have. It was the result of Henry's dipping in at our request to the non-existing reservoir of planned projects in the Biological Survey that [W. T.] McAtee [head of the Division of Food Habit Research] wrote his carefully studied plan for the reorganization of the Biological Survey, with himself as Chief, and a somewhat violent criticism of any plan that might be submitted by a magazine editor who got all his ideas from a capitalistic game hog or a cartoonist who was a notorious critic of the New Deal administration.[5]

The performance of the Biological Survey was judged so poor that committee chairman Beck was in favor of recommending to President Roosevelt that the Survey be abolished because its personnel were neither scientific nor acquainted with the problems of migratory waterfowl management.[6] Darling and Leopold, however, dissented:

Both Leopold and I, knowing that underneath the surface officials of the Bureau there were many competent scientists who if given a chance could be of great service, refused to accept Beck's liquidation recommendation. Tom Beck was at that time president of "More Game Birds" which later was incorporated under the present title of "Ducks Unlimited." Generally speaking, Beck advocated the theory held by the "More Game Birds" crowd—that the way to restore ducks was to hatch them in incubators and turn them loose into the flight lanes, in other words restocking by artificial methods. Leopold and I held to the principle that nature could do the job better than man and advocated restoring the environment necessary to migratory waterfowl, both in the nesting areas, the flight lanes and the wintering grounds.[7]

The Beck committee meetings became so stormy that the threesome could not agree on general policy. Darling eventually "took the whole batch of stuff home and wrote what I hoped would be a compromise between Leopold and Beck" on the question of the Biological Survey. Darling's version of the report emphasized a program of restoration and refuges "for which we had the President's promise of a million dollars."[8]

These meetings, important as they were for the cartoonist caught in the middle, faded briefly from his attention in mid-February. On one of his trips between Iowa and the District of Columbia, Darling was rumored to be "next on gangdom's list of intended kidnap victims." A telegram, bearing an allegedly fictitious name, asked Darling when he would arrive in Chicago and where he would stay. Officers indicated that they had traced the telegram to a Chicago gangster. Federal agents guarded Darling while he stopped in Chicago and escorted him to the

Iowa border, where state officers joined him. Once he had returned to Des Moines, the state agents accompanied him constantly. Darling later said he was flattered that anyone thought he had enough money to be worth kidnaping.[9]

Darling, however, sought funding for restoration programs, not flattery. Restoring the environment, as advocated by Darling and Leopold, was going to require large amounts of money, and similar proposals had earlier been stymied by inadequate financing. Frederic C. Walcott, elected U.S. senator from Connecticut in 1929, led a campaign to create the Senate Special Committee on the Conservation of Wildlife Resources and became its first chairman. (A similar committee in the House of Representatives was headed by A. Willis Robertson of Virginia, who was later a senator.) Walcott was a member of the executive committee of the Boone and Crockett Club and one of the founders of the American Game Protective Association. In 1931, as the waterfowl situation continued to worsen, he pushed for waterfowl management methods to turn the downward curve. His proposals required the infusion of tremendous amounts of new money.[10]

Several suggestions for raising the necessary revenue had been made in earlier years. On the state level, hunting license fees were instituted for the support of restoration programs. It became the prevalent practice of most states, however, ''to throw those monies . . . back into the General Treasury and the conservation Department can have them if they can get them by bills thru the State Legislature.'' Lamented Darling, ''The story is always the same. The natural biological environments have continued on a rapid downward course and the money derived from the harvest of these same resources has been used for every purpose known to man excepting the restoration of wild life.''[11]

Another scheme proposed a tax of one cent per round on shotgun shells and other sporting ammunition. The More Game Birds in America Foundation preferred this approach, while the American Game Association favored a federal hunting stamp, proceeds from the sale of which would be used for waterfowl restoration. Walcott's Senate committee, whose secretary was Carl D. Shoemaker, resolved the dilemma April 4, 1932, when after hearing more than one hundred witnesses it voted in favor of the stamp proposal. Most committee members felt hunters of upland game birds should not be required to help support waterfowl management programs.[12]

As the federal stamp proposal was nearing the end of its laborious route through the Congress on its way to enactment, the Beck committee was trying to draw President Roosevelt's attention to its report. Earl

Bressman, who was scientific advisor to the secretary of agriculture, shared an office with Darling at that time and recalled part of Ding's travails. The Beck committee members had done their work in a matter of a few weeks and had sent a report, unsigned by Darling and Leopold, through the channels to the White House. Darling and his colleagues waited several weeks but heard no response from the president. "Jay . . . was about to give up. He could not understand how their fine report could be so neglected and he was beside himself," Bressman recalled. Finally, Bressman said, the secretary got the president to look at the report and he responded to the effect that it was acceptable. Darling remained convinced more than a decade later that Roosevelt never read the Beck committee report. Ding asserted, "[Marvin] McIntyre, a friend of mine and privileged character around the White House, discovered the Beck Committee Report under the President's bed. At least that's what McIntyre said."[13]

Darling had never been an FDR fan, and the conservationist's first experience in Roosevelt's Washington had left him even more suspicious of what he perceived to be an impenetrable, calcified, bureaucratic maze. At the same time, FDR recognized Ding as a force to be reckoned with—a feisty protagonist, who was not likely to take "no" for an answer, and a cartoonist with a national, front-page showcase for his well-known anti-Roosevelt opinions.

Among the major recommendations made by the Beck committee were: that twelve million of the fifty million acres of submarginal land to be purchased by the federal government be diverted to wildlife purposes, that $25 million be diverted from various programs for wildlife restoration, and that an equal sum be made available from Public Works Administration and other relief monies to hire workers to restore and improve the areas required. Another recommendation, which became part of the report even though Darling and Leopold were opposed to it, called for appointment of a commissioner of restoration, who would work under the direction of a committee of three cabinet members—the secretaries of the interior, agriculture, and commerce. In turn, directors of erosion control, fisheries, wildlife, national parks, and national forests were to function under the direction of the commissioner of restoration.

Secretary Wallace suggested to Roosevelt that the report was "too ambitious to be feasible in the immediate future. I see no possible way of getting $50 million to $75 million for carrying out the plan." Beck had apparently assumed that he had a commitment from Roosevelt and others for at least $25 million, but none of the money was forthcoming until Darling later began to put the pressure on the administration and the Congress.[14] The recommendation for a commissioner of restoration ran into energetic opposition and got nowhere.

His experience as a member of the Beck committee had soured Darling on the performance of the Bureau of Biological Survey and left him disappointed in its leader, Paul Redington. Darling suggested that, "largely because of failing health and partly, I suppose, from bureaucratic habit and fear of criticism and political pressures," Redington was a "totally inarticulate" spokesman for the cause of wildlife restoration. Walter Henderson, assistant chief, and W. T. Mc-Atee, who headed the Division of Food Habit Research, seemed to Darling to be running the Survey but had advanced no program to meet the duck crisis. "Out of fairness to them," Darling later recounted, "we accompanied our Committee Report with a copy of their recommendations produced at our request."

Secretary Wallace read the committee report and the Survey recommendations and, according to Darling, "the contrast was too evident to be ignored." Wallace, in effect, gave Redington an opportunity to take the Beck committee report and put it into effect. For whatever reason, Redington chose not to do so. The Survey was "in the doghouse," both publicly and politically. Redington resigned. Henderson and McAtee fought each other bitterly for Redington's position. Neither, however, was fully acceptable to Secretary Wallace.[15]

The spotlight turned toward Darling. Ding, who for a quarter-century had been a national figure in political cartooning, was about to become a national leader in conservation as well.

# 12

## *The Chief*

WHEN THE BECK COMMITTEE report was finally delivered to Secretary Wallace and President Roosevelt, Darling returned to Des Moines "glad to be relieved from the stresses of the Beck-Leopold violent quarrel. . . ." He had hardly resettled into his daily routine, however, when he received a telephone call from Roosevelt asking him to come to Washington and take over the administration of the Biological Survey.[1]

Darling needed time to consider the matter. FDR was asking a rock-ribbed conservative Republican political cartoonist to join the Roosevelt administration at an influential level, and Darling was not sure he wanted to "aid and abet" the opposition. Roosevelt was also asking this outspoken critic of his administration to forego a six-figure income to become an $8,000-a-year member of the New Deal team. It was an unlikely prospect at first glance. Darling, however, believed he might have something to offer the Biological Survey as a product of his financial independence and political stance. He would have the freedom to shake things up and to take some risks in putting the Survey train back on the tracks. Darling had a first-hand feel for how things were in the nation's fields and waterways. He had a formidable background of getting things done in conservation matters on the state level. He had ideas he believed could benefit the nation if he were given an opportunity to implement them. It was a once-in-a-lifetime opportunity to benefit posterity, and from that perspective it was difficult to ignore.

Darling conferred at length with several persons, including Henry Wallace and Rex Tugwell. They were reassuring. Ding agreed to a temporary assignment on the conditions that he would be given full authority without interference from the "hunting clubs crowd" and that he would immediately be given funds to inaugurate the refuge program. "A million dollars had already been promised by the President," Darling later wrote, "and I got a verbal renewal of that promise the first thing, before I took the oath of office."[2]

McAtee, who had been passed over in Darling's favor, had submitted to Secretary Wallace a lengthy report in which he detailed Darling's "incompetence." Following the cartoonist's appointment, Wallace showed the document to the new chief. "Many of the incidents of my brief experience in the Biological Survey have faded with time," Darling recalled, "but I'll never forget the day when Henry Wallace called me into his office" and displayed the McAtee document. "I never subscribed to the New Deal philosophy and I still wonder at the decision of Rex and Henry to turn down the McAtee report and give me the green light," he wrote.[3]

Darling's appointment was announced March 10, 1934, just four days after Roosevelt signed the new Duck Stamp Bill. Many years later Darling still harbored the suspicion that Roosevelt had appointed him less for what he could do for wildlife conservation than for what silencing Ding could do for the New Deal. Whatever the rationale behind the appointment, conservationists expressed guarded jubilation. They knew a man of stern stuff had been put in charge of a confused situation, and they knew Darling had the backbone required. But was Darling an administrator? Could he cut through the Washington red tape? Was anything really going to change for the improvement of conservation?[4]

By the time he reached Washington, Darling had a mental scheme for reorganizing the Bureau of Biological Survey. He spent his first day as chief going over his proposed changes with Gus Silcox, chief of forestry, another arm of the Department of Agriculture. Darling had been impressed with the forestry chief's sound advice in the midst of the Beck committee investigations. The session ended at midnight. Silcox was in agreement with Darling's proposals. As a matter of fact, Darling reflected, "I suppose the pattern I had drawn up was based as much on his information and advice as my own judgment." Darling dismantled the organization chart. He reshuffled the entire staff. He broke all the rules of Civil Service regulations; he hired new technicians, where necessary, to round out the lopsided divisions.[5]

One of Darling's first moves was to replace McAtee as the head of Food Habits Research. When he showed Wallace his proposals the following morning, "Henry laughed when he saw what we were going to do with the Food Habits Research, and he didn't object." This was not the first time Darling had suggested to Wallace that there were several excellent biologists in that area who were doing all the work while McAtee was taking all the credit. Darling conferred with the

Survey leaders to assess their plans. Nothing resulted from the conferences, according to the new chief, and his patience was "well nigh
exhausted." After giving each member of McAtee's staff a chance to
report, Darling turned to McAtee himself and asked whether he had
any plans that might increase the efficiency of the Survey and his division. McAtee replied that he had ample plans. "Where are they?"
asked Darling. McAtee pointed to his brow. "In here," he said.
"That's a hell of an exhibit to show to the Secretary!" Darling
retorted.[6]

When Darling chose a successor to McAtee, he singled out
Clarence Cottam, a native of Saint George, Utah, who had received
bachelor's and master's degrees in biology from Brigham Young University. He was working on a doctorate, which he received from George
Washington University in Washington, D.C., two years later. He had
joined the Biological Survey in 1929 as a junior biologist. Cottam,
whose career in the Survey was given a substantial boost by Darling's action, later became dean of the College of Biological and Agricultural
Sciences at Brigham Young University and, still later, organizer and
director of the Welder Wildlife Foundation headquartered in Sinton,
Texas.[7]

Darling wrote twenty years later that Cottam "is the most competent, efficient and courageous member of the present staff" except for
J. C. Salyer II, "who should be accorded an equivalent rating." Salyer
was "the one outsider who was brought in and to whom more credit is
due than he will ever get," Darling stated. Without Salyer, Darling
asserted, the success story of the National Wildlife Refuge development
could not have been recorded. There were others, Darling acknowledged, whose contributions during his tenure as chief were substantial,
"but that boy was my salvation."[8]

Salyer was a 1927 graduate of Central College in Fayette, Missouri.
He taught high school science in the public schools in Parsons, Kansas,
from 1927 to 1930. After receiving his master's degree from the University of Michigan in 1931, he became an instructor of biology at Minot,
North Dakota. Darling discovered Salyer when the young ornithologist
participated in the biological survey of Iowa in 1931 and 1932, sponsored by the Iowa Fish and Game Commission. At thirty-two, Salyer
became the first head of the nation's wildlife refuge system, which at
that time consisted of a few unadministered areas set aside for various
species of wildlife. By 1961 when Salyer left the post, the system had
grown to 279 refuges embracing twenty-nine million acres.[9]

In the heat of his supercharged reorganization activity, Darling
discovered that some Survey fieldmen had not seen a representative of

the Washington home office in years. He called for written reports in which fieldmen were asked to suggest areas suitable for restoration. After they had made their reports, they were summoned in groups to Washington for consultation with the new chief. Darling's successor was among them.

When the group of about ten men from the northwestern United States came, among them was a very large pachyderm type of man with decided bucolic attributes, who walked into my office one early morning in 1934. He didn't look like much but when his turn came to speak for his section of the country he immediately justified his position by a very well-stated analysis of Oregon, Washington and the immediate vicinity of his region and told, in detailed terms, the great needs for each specific area and what the possibilities for restoration might be. . . . Naturally, he got immediate attention when it came to the allocation of funds. He set up his restoration program before anybody else had even got started and his choice of personnel among the engineers and construction men was a thing to delight the Chief of the then Biological Survey.[10]

The man was Ira N. Gabrielson, who had joined the Biological Survey in 1915 as an assistant in economic ornithology. Darling brought him to Washington as a consulting specialist. Gabe followed Darling as chief of the Biological Survey and became the first director of the new U.S. Fish and Wildlife Service.[11]

When Darling took office, the Survey claimed just twenty-eight game protectors stretched over the United States and Alaska. There was virtually no federal deterrent to discourage lawbreakers. Darling responded by quietly organizing groups of federally employed agents who could be mobilized to strike swiftly in trouble spots—of which there were plenty. Market hunters were slaughtering waterfowl by the thousands in California and Maryland. Hunters in the Midwest had been shooting ducks in the spring in violation of the Migratory Bird Treaty Act. Darling's phantom force struck in a three-day raid that resulted in forty-nine convictions. One lawbreaker, who had tried to shoot his way out, was seriously wounded. It became suddenly apparent that the Biological Survey was under new management and that the new boss meant business.[12] It was difficult, however, to cover all the bases. Enforcement of the federal law was doubly difficult because the states and their officials were often antagonistic toward the Survey and occasionally conspired with local sportsmen in setting their own seasons and bag limits for ducks in violation of federal regulations. At that time, commercial shooters were marketing their canvasbacks and redheads, killed on Maryland's eastern shore, to nightclubs in Philadelphia, Baltimore, Chicago, and New York. Without money for additional enforcement officers Darling felt his hands were bound. But Darling knew how to get his story told in the press. He had complained publicly about Roosevelt's foot-dragging on Survey appropriations, and he complained just as energetically of the widespread violations the Survey seemed powerless to stop. Darling's cries were heard by a man "whom I had never known and only heard of as an economic royalist"—Richard Reynolds of the Reynolds Tobacco Company.

There appeared in Darling's office a strange man in military uniform "who stuttered so badly that I could not understand him and it was very difficult to get any idea of what he was trying to tell me[,] but he laid on my desk a check for $2,500 signed by Richard Reynolds." Darling was in a quandary. The check made out to Darling was a poker kitty dedicated—by agreement among the players—to enforcement of game laws on Maryland's shore. Darling saw no way personally to use the check for government activities because of legal complications. The mysterious man in uniform, however, became the solution to the problem. The visitor was an amateur taxidermist and interested in wildlife, had nothing else to do, and needed little income for his personal support.

He volunteered to go down on the eastern shore of Maryland, equipped as a taxidermist with a license from the federal government and the state of Maryland, to shoot or trap birds for scientific purposes for the Smithsonian Institute. By living in a tent and with the accessories of a Model-T Ford, he invaded that area and worked for a long time without any reference to law enforcement or any attempt to even acquaint himself with the local population. In fact, he went alone and unattended, but in the course of time he made the acquaintance of some of the men who were engaged in shooting ducks for the market. They were all poor folks and we weren't interested in making them the goats—what we wanted was the key to the commercial burglars who were hiring these fellows and paying them two bits per duck and selling them to the night clubs.

The "Maryland connection" eventually resulted in the location of a man who was shipping ducks under the name of his former partner, whom he had murdered and buried in the swamp. The duck buyer went to prison for life for his partner's murder, and through the cooperation of Reynolds and his $2,500 check the market shooters of eastern Maryland were put out of business.[13]

But the new chief still needed money, and he was having trouble getting it. The $1 million promised by FDR sounded to Darling "like an awful lot of money."

It was needed badly if we were to even begin our program[,] for the Bureau in 1935 had been cut to the bone on its budget. The President never made good on his promise, however, and while I was busy reorganizing the Bureau staff to implement the restoration program I sought means of financing other than the million promised by the President, it having become quite evident that he did not intend to make any money available. I tried Harry Hopkins, who had the major relief appropriations under his control. He studied our project program and in the presence of Wallace promised me six million, and then went to Europe and never ga[v]e me a nickel.[14]

A Darling acquaintance remembered those frustrating days in Washington and an incident involving Darling and Hopkins, the ambitious and dedicated "social engineer." Hopkins—as head of the Federal Emergency Relief Administration, the Civil Works Administration, and the Works Progress Administration—handled billions of dollars in the New Deal era. A skillful manager, he could be arrogant and insistent on placing human needs at the top of his long list of priorities. With a background in social work, Hopkins was opposed to what he defined as socialism for the rich coupled with demands for free enterprise for the poor. He was the personification of the New Deal whose philosophies Darling detested and distrusted. Hopkins also hailed from Sioux City, where he had been born forty-five years earlier.[15]

Dr. Earl Bressman, Wallace's scientific adviser, recalled that Jay had come to Washington full of hopes and dreams of what could be done in regard to wildlife and that he felt rebuffed at nearly every juncture. One day, Secretary Wallace asked Bressman to come into his office. "Jay was there," Bressman recounted, "red-faced and upset. He had told Wallace his troubles." Wallace ordered Bressman to make a high-priority item of funding for the Survey, to do all that was possible to obtain the necessary financing, and to call on him for any contribution the secretary of agriculture could make to the effort.

Bressman suggested that Harry Hopkins would have to approve Darling's request for funds to purchase land. Wallace and Secretary of Interior Harold Ickes had already approved the appropriation, Bressman said. At Bressman's request, Wallace arranged a meeting the following morning for Bressman, Darling, and Hopkins. When Darling and Bressman arrived, Hopkins was seated in his office, surrounded by a platoon of his assistants. "Mr. Hopkins, who had known very well, asked what he could do for us after he introduced us to his assistants," Bressman recalled. The scientific adviser then launched into a presentation, spelling out Darling's difficulties and how they could be dispelled if Hopkins would approve the Survey's request for funds for land to be set aside as refuge areas.

At the conclusion of the Bressman presentation, Hopkins responded, "I don't know if we're interested in the relief of birds." Darling, silent to that point, jumped to his feet and shouted in Hopkins's face, "Harry, I was a trustee at Grinnell College when you were just a student!" Darling also charged that when he was considering the Survey appointment Hopkins had strongly encouraged him to accept it. Hopkins interrupted the chief. "Where do I sign these papers?" he asked. "See," Hopkins added, turning to his assistants, "that's how you get things done in Washington."

The Resettlement Administration promised to use some of its funds to buy up distressed farmland in regions where refuges were needed, Darling asserted, but never followed through. "Ducks seemed unimportant to any of the authorities," he complained. The Duck Stamp Act shone as a light at the end of the tunnel. It provided for the sale of a federal stamp to every hunter of migratory waterfowl. The proceeds from the stamp sales were to be earmarked for refuges and their management and for enforcement of migratory bird regulations. As one of his early obligations Darling agreed to do the artwork for the new duck stamp. That illustration, a vigorous likeness of two ducks about to descend to the water, became one of Darling's best known works and a collector's item sought by sportsmen throughout the nation. It was done, Darling recounted, from a preliminary sketch executed by him in preparation for a luncheon meeting with the head of the postage stamp engraving department. The sketch was one of several prospective designs Darling had taken to the meeting. He was later chagrined to learn that the Bureau of Engraving had unilaterally made its choice of designs and had set the presses in motion.[16]

Although the Duck Stamp Act had been passed, it had not been funded. Darling, desperate and out of patience, was about to try another tack. "I am completely oblivious to anyone's slant on FDR's contribution to the Duck Restoration Program," Darling later wrote, "but so far as I am concerned he blocked me, and consciously too, in every effort to finance the program which he himself had asked me to carry out." Whenever Darling complained to the president that he had not received the $1 million he had been promised, FDR would write out an I.O.U. for $1,000,000 and tell Darling to hand it to Harry Hopkins. Hopkins, recognizing the president's humor at work, would ignore the chit and suggest that Darling make his pleas to Ickes or Wallace.[17]

Darling finally sought aid "over on the hill, which resulted in one of the funniest incidents in the whole restoration procedure." It was late June or early July of 1934 and the restoration, which had been well detailed by that time, remained stalemated for lack of resources. One of the few members of Congress who fully understood what Darling meant when he said, "ducks can't nest on a picket fence," was Peter Norbeck, a senator from South Dakota. Norbeck, who had a strong Scandinavian accent, agreed to ask for unanimous consent for a Senate resolution giving the Survey $1 million from unexpended relief funds carried over from the previous year. "There was plenty of that I knew," Darling asserted. He described the events that followed:

Senator Norbeck agreed to put it through. When he got to the floor of the Senate, he got a new idea. The Duck Stamp Act was then up for final passage in the Senate. Norbeck rose to speak on the Duck Stamp Bill. He removed his false teeth and asked, in words totally devoid of understandable articulation, for unanimous consent for an amendment to the Duck Stamp Bill allocating six million of any unexpended 1934 relief funds for the Biological Survey restoration program. It passed unanimously by voice vote and the Senator engineered it through the House-Senate conference committee the same afternoon.

Roosevelt was scheduled to leave the following morning for a Caribbean fishing trip. Darling had urged FDR to keep an eye out for the Duck Stamp Bill and to sign it, so that its provisions could be put into effect in time for the fall shooting season.

By special messenger the Duck Stamp Bill, including Pete Norbeck's six million-dollar amendment, was rushed to the White House. He [FDR] recognized it and signed it without reading it, I guess, for when the President returned, after his fishing trip, and found that he had authorized six million dollars for our restoration program he wrote me a letter which I still preserve as one of the most interesting documents I ever received in my career.[18]

That letter came in 1935, after Darling had written Roosevelt a cartoon-illustrated letter urging him to reinstate a $4-million appropriation that, according to Rex Tugwell, the president was going to disapprove. Darling claimed the Biological Survey could make better use of retired agricultural land than anybody. He wrote:

Others just grow grass and trees on it. We grow grass, trees, marshes, lakes, ducks, geese, furbearers, impounded water and recreation.
    The six million we got from Congress and which you think is enough, is mostly going to buy Okefenokee [Wildlife Refuge], the ranches on the winter elk range in Jackson Hole, the private lands that lie in the midst of the Hart Mt. antelope range, and for rehabilitation (dams and dikes) of the duck ranges we bought last year.
    By the way, Secretary Ickes wants me to give him Okefenokee. Do you mind? I don't, only that it cuts into our nesting area funds.
    I need $4,000,000 for duck lands this year and the same bill which gave us

the $6,000,000 specifically stated that at your discretion you could allocate from the $4,800,000,000 money for migratory waterfowl restoration.

We did a good job last year. Why cut us off now?[19]

In reply Roosevelt wrote that he had mentioned to the acting director of the budget "the other day" that "this fellow Darling is the only man in history who got an appropriation through Congress, past the Budget and signed by the President without anybody realizing that the Treasury had been raided." Roosevelt did not stop there:

You hold an all-time record. In addition to the six million dollars ($6,000,000) you got, the Federal Courts say that the United States Government has a perfect constitutional right to condemn millions of acres for the welfare, health and happiness of ducks, geese, sandpipers, owls and wrens, but has no constitutional right to condemn a few old tenaments [sic] in the slums for the health and happiness of the little boys and girls who will be our citizens of the next generation!

Nevertheless, more power to your arm! Go ahead with the six million dollars ($6,000,000) and talk with me about a month hence in regard to additional lands, if I have any more money left.[20]

One way or another, however, Darling and the Biological Survey were organized and fairly well funded; and the formerly lethargic agency would never be the same again.

# 13

## *Train on the Tracks*

DARLING and his Bureau of Biological Survey had broken the funding logjam. Even so, things were not as good as a U.S. Fish and Wildlife Service pamphlet later described them as being. The pamphlet read:

In a short time, $8,500,000 of emergency funds were obtained to buy lands and construct fences, dikes, dams and necessary buildings as follows: A special fund of $1,000,000 was set aside by the President for the purchase of migratory waterfowl refuges; $1,500,000 was allocated from the submarginal land retirement fund; $3,500,000 was allocated from drought-relief funds, for purchase and development of lands within drought-stricken areas; $2,500,000 was allotted from WPA funds, for engineering operations, to construct water-level controls and to improve the refuges.[1]

The chief wrote in rebuttal:

I am peculiarly interested in the statement that $1,500,000 was allocated from the submarginal land retirement fund, $3,500,000 from the drought relief fund and $2,500,000 from the WPA fund. Those figures, which sound as though they were cash funds turned over to the Biological Survey, have no place in my recollection.

Darling admitted that some such allocations may have come to the Survey following his departure, "but during my day all we got from them were CCC camps, some WPA gangs of workers and a more or less friendly attitude on the part of the resettlement administration. . . ." Most of what the Survey received, he asserted, it had to steal, "with the cooperation of some of the subordinate administrators" while the top-level executives "turned down every request we made for cash allocations."[2]

But Ding finally had some cash in his Survey pocket and he had ideas to go with it. One of the first was aimed at stemming the continued decimation of the waterfowl flights. As duck populations kept dropping, Darling in 1935 clamped down with the tightest hunting restrictions in history.[3] The open season was trimmed to thirty consecutive days; bag limits were reduced to ten ducks and four geese; bait was outlawed, as were live decoys; shotguns holding more than three shots were made illegal. Darling recalled the reaction:

The rigid restrictions which were put upon the shooters at that time were pretty violent and caused a good deal of criticism. . . . The answer to all our critics was that we knew very well how to kill more ducks but that was not the problem; that until we could restore the nesting grounds and have better duck factories to produce more ducks, they would have to get along with smaller bag limits and shorter seasons. That worked pretty well when the boys finally realized what the crisis in the duck population was.[4]

Congressmen whose constituents opposed the clampdown assailed Darling in droves. Ding met the assault head on. "The regulations," he told them, in effect, "will stay as long as they are needed to bring back the ducks; and if tougher restrictions will help the cause, we'll find some tougher restrictions."[5]

Meanwhile, Ding was eager to put the Cooperative Wildlife Research Unit idea, operated successfully for three years at Iowa State College, into operation on a broadened scale. He put together a proposal that would include nine land-grant colleges at a cost of $243,000 for three years. He had managed to get the colleges and their respective state conservation departments to pledge two-thirds of that amount, but he needed the remaining $81,000. It again became obvious that the money was not going to come from the Roosevelt administration.[6]

Darling turned to a group of industrialists with a collective vested interest to help him out. At a dinner meeting at the Waldorf-Astoria Hotel in New York City April 24, 1935, Darling made his plea for help. As a result, the Du Pont Company, the Hercules Powder Company, and the Remington Arms Company—urged by C. K. Davis, the head of Remington—agreed to subscribe funds to federate the organized sportsmen of North America, to stage a national wildlife conference in Washington, and to help underwrite the proposed wildife training courses at the land-grant colleges. Darling commented, "Out of this single meeting there emerged, either directly or indirectly, the Cooperative Wildlife Research Unit Program, the American Wildlife Institute, the National Wildlife Federation, and the North American Wildlife Conference—a rather productive three hours or so of dinner conversation!"[7]

Nearly twenty years later, Darling recalled the moment Davis came to his aid, and he contemplated the consequences if Davis had not acted:

It was pleasant to have you recall that memorable dinner in New York back in 1935 when good old C. K. Davis got up on his hind legs and gave the conservation program a shot in the arm that sent blood tingling through the veins, beginning with me, and which put pink in the cheeks of everyone in the wildlife game clear across the American continent. My God, just think of what a transformation took place in the broad expansion and methods of wildlife management as a result of that "command decision!" And then think for a minute of what could have happened to our wildlife conservation program if "C K" hadn't spoken just as he did, and just at that particular time. I don't suppose there are many people in the world, not even among those who are most intimate with the workings of our conservation activities, who realize by what a narrow thread our whole restoration program was hanging.[8]

To Davis himself, Darling wrote:

I never will forget that day when I presented the case of the Cooperative Units to the group of executives representing the Sporting Arms and Ammunition Manufacturers. I can still see the circle of faces and I could read little encouragement in their expression until you, "C K," who I didn't know then as well as I learned to know you later, arose to speak. Frankly at that moment I was scared to death. It was the culmination of many weeks spent in visits and correspondence with the State Conservation Commissioners, Deans and Presidents of Land Grant Colleges, and arguments with the Secretary of Agriculture and even the President of the United States on the wisdom of the procedure. If I couldn't get the support of the Sporting Arms and Ammunition Manufacturers my whole house of cards was bound to collapse.[9]

Davis promised Darling, "We are going to back you" even if Remington had to "go it alone." Darling was impressed with the hands-off attitude of the firearms and ammunition makers. They "don't do much dictating, or at least, I never found it so," he wrote. "They gave me considerable financial support while I was in the Biological Survey and tied no strings whatsoever to their contributions."[10]

With the promised support the Cooperative Wildlife Research Unit concept was expanded to include ten states. Since that time the program, expanded to far more colleges and universities, "has produced an amazing volume of original information on wildlife problems and has developed scores of new techniques in wildlife management while training literally thousands of young men for professional careers in wildlife work."[11]

Darling was also influential in the passage of the Pittman-Robertson Act, designed in part to provide public relations activities on behalf of conservation—a need Darling the newspaperman had recognized for years. Such efforts were to be financed by the excise tax on sporting firearms, shells, and cartridges; such funds were to be earmarked for conservation purposes and were not to be diverted to any other use.[12] The old cent-a-shell scheme, which had been devised to provide the restoration money the Roosevelt administration seemed unwilling to provide, failed to generate sufficient funds. When the "nuisance taxes" were devised to aid economic recovery, the proposal for a tax on sporting ammunition was raised as justification for a specific appropriation for wildlife restoration. Congress did not approve.

Darling got Roosevelt to agree that taxes collected as a result of the use of natural resources should be plowed back into restoration. Writing to a friend, he said, "The President . . . said he thought it ought to be done now and that if I would secure a written agreement from the ammunition makers and sporting arms manufacturers to turn over . . . ten percent for use in restoration projects he would take off the nuisance tax on sporting arms and ammunition just as soon as recovery had been accomplished."

As much to his own surprise as anyone's, Darling obtained a written agreement from every arms and ammunition maker in the United States to pay ten percent of gross receipts into federal conservation channels. And, wrote Darling, there was more:

They not only would agree to pay ten percent of gross receipts into the Biological Survey but would continue it for a period of five years after the nuisance tax was withdrawn. Of course the President didn't think I could get any such agreement and when it arrived, all signed and documented, he referred the matter to the Secretary of the Treasury who reported adversely. That was in 1935, after the depression was pretty well washed out and many of the nuisance taxes had been eliminated.[13]

One of Darling's proudest accomplishments during his term as chief of the Biological Survey was the nearly miraculous restoration of the Sheldon Antelope Refuge in Wyoming. His earliest concern for the area was sparked by Theodore Roosevelt many years earlier, when TR was co-editor of *The Outlook* magazine. Darling had sat in Roosevelt's office as the Rough Rider related his fears concerning the disappearance of the pronghorned antelope and how the Boone and Crockett Club had purchased land for a permanent refuge and turned it over to the U.S. government for administration.[14]

By the time Darling became Survey chief, the Sheldon Antelope Refuge had been "so mistreated that even a grasshopper would starve to death.'' Cattle ranchers and sheep herders had grazed the area to bare dirt and moved on when there was no food left for their animals. ''There weren't enough sage hens left to tempt a hungry coyote and the few hardy prong-horned antelope which remained on that area were approaching extinction and none of their progeny lived over the next winter after they were born.'' When Darling visited the area, he was stunned. The refuge had never been fenced. Ranchers and herders turned their stock loose on the grounds at will. The seven waterholes Darling had viewed years before were ''dry as a baked 'dobe brick, the land as barren as the middle of the paved thoroughfare of Broadway, New York.'' Darling judged it ''the most desolate piece of the American continent I had ever visited.'' It was a federal refuge, though, and the chief was obliged to make it perform as one.[15]

Ding was not certain of his course of action. His only immediate thought was to put a fence around the place to keep cattle and sheep from scouring the earth. That single step was the first in a remarkable cycle that taught even the ecologist Darling some priceless lessons, and it was the beginning of a demonstration project that he often displayed in his ecological missionary work in the years that followed. The fence kept the cattle and sheep out, and by the fall of the same year there was enough ryegrass and bunchgrass growing that the winter's snows were caught and held there ''in one great white sheet of landscape.'' Outside Darling's fence, however, the grass had been gnawed away to the roots, and the snow had blown off into dry gullies and left fields of dusty wasteland.

The following spring, the snow in the gullies melted and ran off through the ditches, carrying valuable soil with it; but on the refuge the snows melted and soaked into the ground rather than running off. ''That summer,'' recalled the fence builder, ''there was water in those seven water-holes up until the middle of July and that was the first time it had happened in many years.'' The second summer the water holes were full at the end of August. Several years later, vegetation covered the entire refuge, and the area benefited from a continual and sufficient supply of water. An old well on the site, long dry, was being pumped regularly and was fed entirely by the ground storage of winter precipitation, Darling reported.[16]

The chief was amazed at what a little ryegrass could do for a virtual desert. The man in charge of the Sheldon refuge continually pleaded with Darling for more money, ''which he needed, all right, because somebody kept cutting the wire fence.'' But when the manager, at the end of the second year, ''prayed for money to cut fire lanes through the grass that had grown up since we fenced out the domestic herds—that was something I had difficulty believing until after I went out and had a look."[17]

Darling was not successful every time he climbed into the ring. He opposed huge dams in general, contending that the place to retain and control precipitation was on the land where the moisture fell. He saw little logic in letting the rush of water grow so large and violent that only ecologically disastrous tourniquets on the nation's largest waterways could avoid tragic flooding. He winced at the thought of priceless topsoil being washed away by equally priceless water, only to build up behind dams as useless, sterile silt. He resisted the intentional flooding of rich bottomlands behind the dams. He had long been an outspoken critic of the U.S. Army Corps of Engineers, who proposed and built the dams Darling believed were strangling the nation's rivers.

He lost the battle in the fight over the Santee-Cooper Dam, but he won a staunch ally in Secretary of the Interior Harold Ickes. Relations between Darling and Ickes grew mutually rancorous, however, before they became mutually respectful. In a routine memorandum Darling noticed that the sum of $51 million in relief funds was being set aside to build a huge dam across the Santee River Valley in South Carolina, thus diverting the water into the Cooper River Valley. Darling sent a team of his scientists to the scene to determine what the effects of such a dam might be. The report said the dam would cut off the freshwater inundation from the last large virgin stand of tidewater cypress trees on the North American continent. It said the dam would seriously affect the ecology of the waters back of Cape Romaine, the location of one of the largest shrimp shoals then known and the spawning grounds of several species of fish, clams, and edible crabs. The $51 million price tag for the dam was small when compared with the loss in timber and natural resources.

Darling noted also that "once the ecology was destroyed, it was destroyed forever, so I let out a blast in the newspapers against this project, and against Mr. Ickes," who was supposed to be in control of Public Works Administration money. The Secretary of the Interior's response was vehement. According to Darling, "Next day Ickes let me have both barrels. He had never heard of the Santee-Cooper project, and none of his money was going to build the Santee-Cooper dam and power project, and I was a vicious and conscious liar, wilfully trying to destroy his effectiveness as a government agent, and I was a dirty little Republican anyway!"

Darling took a copy of the memorandum, in which he had read of the $51 million fund, to the secretary's office. "Ickes hit the ceiling and immediately got F.D.R. on the telephone while I was sitting in his office," Darling recalled. The president apparently confirmed the contents of the memo, and "from that moment Ickes was on my side." The pair rushed to the White House to explain to Roosevelt the scientifically determined results of constructing such a dam. The president "was very much astonished to find that the dam would have such destructive consequences and assured me that he would withdraw the project and the appropriation," Darling stated.

Two weeks later, according to Darling, the president called him to say he had a report from the South Carolina State Conservation Commission refuting the Biological Survey's findings. James (Jimmy) Byrnes, who later became secretary of state under President Harry Truman, was then running for the senate from South Carolina and had managed to get his state Conservation Commission, a purely political body, to write its own version of the Santee-Cooper project, Darling charged. "It was wholly fallacious and pure fiction, but the dam was built," the dejected Darling noted. His dejection was intensified when in 1951 the former chief stopped to see what had happened to the cypress forest. It was a "ghastly graveyard" of bleaching limbs and trunks of a once magnificent stand of timber, he wrote. Even so, Darling was impressed with Ickes and the response to his pleas for help. "Ickes fought this Santee-Cooper project with me with every vindictive violent expression that he had in his body, after he knew what the highly qualified scientists had to say about the consequences of it," Darling asserted.[18]

Darling projects during his twenty months as chief of the Biological Survey were too many and diverse to catalog readily. In each of his many frays—whether with the Bureau of Reclamation over supplying water to the Tule Lake Federal Refuge in Oregon or with the Corps of Engineers over incorporating improved fishways in the Bonneville Dam on the Columbia River—Darling was a disciple of the scientific method.[19] He insisted on giving his highly trained specialists in the life sciences freedom to expose unvarnished scientific facts and let the chips fall where they would. At the same time, he clearly saw a need for incorporating conservation education in the lives of Americans at several levels.

Finally, as part of the effort to educate the population in what Darling called the oldest legal code in the world, "The Laws of Nature," he saw an essential role for a centralized source of conservation information. He envisioned a conservation clearing house maintained by the many and burgeoning groups with interests in one or another facet of conservation. Because so many groups were interested in only limited aspects of conservation, he suggested, it was difficult for the population in general to see the ecological picture; it was not clear to most fighters in the conservation trenches that their efforts were related and interdependent and critical to the health of the nation. Darling saw a need for an intelligence unit for this army of conservationists and for the population as a whole. He saw nothing more important to be accomplished than this herculean educational task. Long before the word "interdisciplinary" became jargon in formal education, Darling saw the wisdom of folding ecological education into a variety of disciplines; tying them together with an interrelated thread. It was just such an approach to biology that had kindled his interest in his days at Beloit College.[20]

At the Waldorf-Astoria dinner meeting of April 24, 1935, Darling argued for a federation of the hundreds of local, county, and state organizations devoted to wildlife conservation. He said the approximately 6,000 such organizations then in existence were handicapped for lack of a central coordinating agency at the national level. One result was the incorporation of the American Wildlife Institute in Washington, D.C., July 22, 1935, with Walter P. Chrysler as chairman of the board and Thomas H. Beck as president.[21] Several months following Darling's resignation from the Biological Survey, the American Wildlife Institute sponsored the North American Wildlife Conference. That meeting led to the creation of the National Wildlife Federation, with Darling as the organization's first president. It was another significant outgrowth of Darling's tour of duty as chief of the Survey.

Darling's tenure was drawing to a close. He had, after all, accepted the appointment on a temporary basis. He believed his free-swinging style, while it was helpful in getting the agency moving again, was also beginning to alienate some persons upon whom the Biological Survey was heavily dependent. His relationship with President Roosevelt had also been stretched until it had sprung. Darling had been overruled on a plan to reflood the James River Valley in South Dakota with water that was eventually trapped farther upriver behind the Fort Peck Dam. "That was the bitterest defeat I suffered during my brief period of alleged authority," he wrote much later, "and whenever I hear anyone boasting of Franklin D. Roosevelt as a conservationist I think of how little the Public knows of the political crimes committed in the name of Conservation."[22] Finally, his body was having trouble keeping up with his spirit. He was nearly sixty and working incessantly.

Darling told Wallace he was ready to resign and, at Wallace's request, took a hand in selecting his successor. "I think no one was more astonished than Gabrielson himself when I called him in one day . . . took him in to the Secretary of Agriculture and introduced him as the man who should be made Chief of the Biological Survey,'' Darling wrote.[23]

A week before his resignation became effective, Darling retired to his room at the Mayflower Hotel to compose a difficult letter to Clark Salyer, his young and able recruit to government service:

The jig is up and I am to leave the Survey November 15th—with my tail between my legs! I am to be succeeded by Gabrielson as chief. The papers are all signed and sealed and I am waiting with such patience as I possess until the zero hour of official announcement. Please say no word about it until the Secretary makes the fact known.

It came about in this way. My engine got overheated and my valves began to leak. An examination in Rochester left me only one choice, to either get out or slow down on the job to a snail's pace. I tried the latter and can't do it. Added to that Mrs. Darling will not live in Washington. I have trod on so many sore corns in the Government circles that it is beginning to reflect on the functioning of the Bureau—seriously. And there are 999 other reasons just as good.

You will probably feel that I have been false to my promises to you about my successor in office. I probably am entirely responsible for Gabrielson's selection instead of yours, although the final choice rested with the Secretary.

It came about as follows. I put your qualifications—all of them, which included the most powerful personality and the most vigorous leadership and technical qualifications with one black mark against you, i.e. your ability to think of more caustic things to say than anyone in the world and *saying* them instead of keeping some of them to yourself—up to the Secretary and set over against them the qualifications of Gabrielson who has many of your traits with a muffler on his disposition.

Darling also noted that Gabrielson had twenty years of service with the Survey, while Salyer was a relatively new recruit. Gabe's advancement would reestablish the principle of promotion for efficient service "which was jettisoned when I was brought in and stuck on top of 2,400 who had devoted a lifetime to the work."

Darling was convinced that Wallace made his decision largely on the basis of Gabe's longer service, "but there was also something of the feeling lurking in the back of his head I guess that he's had about all the temperamental leadership he could stand for a little while." Darling claimed that the president, the Army, and some citizens with considerable political influence were eager to see him leave. "Anyway," he wrote Salyer, "it's been a great war. I've done the best I knew how and had it not been for you I could not have done it at all."[24]

Secretary Ickes, who had shown an interest in getting the Biological Survey and the Forestry Service transferred from the Department of Agriculture to the Department of the Interior, wrote Darling: "I don't like it at all that you have resigned as Chief of the Biological Survey. I feel a distinct loss in your going. You had been doing a fine job, and now that you are no longer head of the Survey, I haven't nearly the interest that I did have in attaching it to the Interior Department.[25]

When Darling's resignation was announced, his departure was mourned in editorials from New York to Los Angeles. The newspapers were undoubtedly going to miss the outspoken and colorful chief, the man who would as soon say something outrageous to a reporter as confer first with the secretary of agriculture, the man who knew the demands of the newspaper business as well as those of conservation. "My place in the picture came about by pure accident rather than by design," Darling explained. He believed that any other reasonably well-informed person, freed of fears for his government job security, "could have done as well and probably better than I did." He contended throughout the years following his brief but active Survey career that, by "telling everybody to go to hell," he was able to put the fine group of eager, scientific men in the old Biological Survey into constructive action.[26]

Years later, Darling expressed satisfaction with his decision to leave the Survey when he did: "What is it about the Washington atmosphere that makes a man, after brief exposure, unable to tell the truth? It's lucky that my stay was short."[27] Darling believed he had accomplished what he could at the Biological Survey. He wanted to return to his successful cartooning career. He also wanted to follow through on the idea of a national federation of agencies committed to conservation. He was firmly convinced that, properly constructed, such an organization could make conservation a national watchword and a way of life.

# 14

## Back to the Drawing Board

"I'VE ALWAYS been given a lot more credit for that rejuvenation of the Bureau of Biological Survey than I deserved," Darling insisted. There were a lot of good men in the Bureau and for the most part all they needed was a chance to do their stuff, he claimed, "and as soon as I thought I had the machine well greased and all the wheels on the rails and going places I came back home and started again to draw cartoons."[1]

It did not take Ding long to get back into the rhythm of daily commentary on matters national and international. Even before his resignation as chief of the Biological Survey, he drew a cartoon titled, "The Fates Are Funny that Way." Individual panels showed persons being killed in automobile accidents and earthquakes and by lightning and poisonous food. At the end John Public said to Mrs. Public, "But nothing ever seems to happen to Huey Long." The fates were not so funny. Two days later Long was shot as he walked the halls of the Louisiana capitol.[2]

Darling, at fifty-eight years of age and with about 175 pounds on a six-foot frame, was the picture of health in spite of his chronic illnesses. An Associated Press writer described Darling in 1935 as a man who "smokes cigarettes incessantly, can swear eloquently and vigorously on occasion . . . collects fine books and is an expert amateur gardener."[3] His work schedule also belied his uncertain health. His correspondence alone kept his secretary fully occupied. Merle Houts, who had married Harry Strasser in 1935 four years after being hired to do Darling's typing, remained Darling's personally paid employee. She was Ding's receptionist; she occasionally answered and placed his phone calls, kept his appointment book, transcribed his correspondence, and did his filing. She sometimes had assistance, but for thirty-one unbroken years she was Darling's loyal aide.

Darling needed a secretary "with four Ph.D.'s," according to a co-worker. Merle did her best and in extraordinary volume, but Darling occasionally became visibly upset with her. Merle said her boss was interesting to work for, and that she never knew what to expect. She claimed he was more patient than others might have been. She had some reticence about working for a compulsive perfectionist. "I had heard that he blew up at other secretaries, before me, and that he had had them in tears. I was never in tears," she asserted.

Darling had a variety of interests and a remarkable reserve of energy to pursue them. He would arrive in the office about 10:00 A.M. and would work straight through the day until about 6:00 or 6:30 P.M. He did not generally go out for lunch unless he had an appointment. "He would finish a cartoon and then he'd ask, 'How much time do I have?' " Strasser recounted. "I'd tell him and he'd put the cartoon aside and start to redo the entire thing." Kenneth MacDonald, who was news editor at the *Register* in those days, was repeatedly driven to distress by Darling's disregard for newsroom deadlines.[4]

Darling was fifty years old and a *Register* veteran when Mac-Donald, twenty and fresh out of the University of Iowa, took a part-time position on the paper in 1926. A native of Jefferson, Iowa, the neophyte had grown up on Darling's cartoons in the *Register.* Mac-Donald remained at the *Register* for a half-century, retiring in 1977 as editorial chairman; but in those early years he saw little of the venerable Ding. The younger man began working directly with the cartoonist only after becoming managing editor.[5] Darling was frequently a problem, MacDonald confirmed. Ding would frantically rework a late cartoon, the perspiration running off his face. When he had finished, he would go home and often read until 1:00 or 2:00 A.M.

MacDonald acknowledged that Darling was quick to anger and that his perfectionism was at the heart of his explosive nature. "He was very impatient with those who didn't share his desire for perfection," MacDonald noted, "and he would spend hours on a small detail of a cartoon. He worked all the time and expected others to do the same." MacDonald described Darling as a "very vibrant character; extroverted, outgoing and ebullient. He had a strong personality. He was quick and impatient of any delays." He was also jealous of the position of his cartoons in the *Register,* and the editorial staff knew it. "He took enormous pride, like any good workman, in what he was doing," Mac-Donald explained. In editorial conferences it was occasionally suggested that an exceptional photograph be centered at the top of the front page and that the daily cartoon be moved down or to an inside page. Invariably, the next consideration was whether it was Darling's cartoon that day or one drawn by Tom Carlisle, his assistant. Darling "didn't like it one bit" when his cartoon was moved, MacDonald recalled. Ding believed he had the equivalent of squatter's rights on the top center of the front page; he felt he had a proprietary interest in that part of the page, where his readers had for years been accustomed to finding his cartoons. Any editor who thought otherwise soon felt the heat of Ding's anger.

Darling had clout. He was not involved in the daily editorial conferences. Instead, as had been the case since the day he was hired by Gardner Cowles, he went his own way. Gardner (Mike) Cowles, Jr., recalled that Darling never wanted to get into the management of the newspapers. He did not interfere in the general operation of the *Register* and *Tribune,* and in return he wanted no interference from anyone where his cartooning was concerned.

In 1935 Darling was a stockholder and officer in the Register and Tribune Company, an advantage he had not enjoyed in 1906. As had also been the case in the past, his views, as expressed in his front-page cartoons, were frequently opposed to those printed on the editorial pages. "On occasion," MacDonald acknowledged, "a cartoon might have been held out." Darling's secretary recalled that her boss "was pretty much king. Nobody questioned him." Nobody questioned him, at least, without giving due consideration to the consequences.

Darling's pique would suddenly bubble to the surface and the feisty Ding would react immediately and instinctively. When he and MacDonald were once entering the same restaurant, but with different parties, Darling sidetracked MacDonald to tell the editor that a conservation story in the *Register* was slanted and uninformed. MacDonald, impatient to rejoin his party, defended the reporter and suggested that Darling did not have all the facts. Darling said that if he was wrong, he would appreciate it if MacDonald would show him where. MacDonald said the topic was complex and that he would gladly call Ding later in the day to discuss it. The afternoon became especially hectic, and in the rush MacDonald failed to call Darling. A sketch from the cartoonist's drawing board was soon delivered to the editor's desk. It pictured Ding with his head on a rock and MacDonald, standing over Ding's kneeling form, about to bring a huge club smashing down. Ding was saying, "Hurry up. The suspense is killing me!" That incident typified to MacDonald the fact that Darling's quick temper, ready criticism, and pugnacity were moderated by the release valve of his rich humor.

Darling, for all the inconvenience his fame brought him, remained very approachable, according to a fellow worker. At the *Register,* the man who shoveled coal in the basement would say, "I think I'll go up and talk to Ding." The cartoonist could be as much the coal shoveler's friend as he was the friend of Gardner Cowles. If Darling was aggressive one minute, he was likely to be very gentle the next.

Gordon Meaney, a barber who had become interested in etching through Darling's example, found the cartoonist less than hospitable, however, at a meeting in Ding's office. Meaney occasionally cut Darling's hair and had seen and admired some intaglio prints Ding had done. "He was hard to get acquainted with because everyone wanted to get close to him," Meaney recalled. One day, "I got the bit in my teeth and went to see him." When he was finally admitted to Ding's lair, the barber blurted, "Mr. Darling, I would love to have one of those etchings." Darling, apparently annoyed at the intrusion, exploded. "I'm not doing that anymore! I'm through with that! I quit!"

When Darling learned that Meaney had a son he was trying to put through medical school, it marked the beginning of a long relationship. When Meaney opened his own shop across the street from the Register and Tribune Building, "damned if he didn't start coming to my shop to get his hair cut," the barber recalled. After his memorable encounter some years before when he had asked for one of Ding's etchings, Meaney had begun a serious study of etching and engraving and finally made a print, which he hung in his barber shop. When Darling spotted Meaney's etching, made with washing machine rollers for a press, the cartoonist yelled, "What is that?" Meaney replied, "That's my artwork. You wouldn't give me any of yours. Would you like one of mine?" "Hell no," Darling said. "You need some tutoring."

Darling had a favorite retreat west of Des Moines at his Peony Farm, where there was a comfortable house with a bedroom and sitting room on the ground floor and etching materials and equipment in the basement. The Darling workshop was well supplied with stacks of fine imported papers, copper plates, and scores of etching tools as well as Ding's large print-making press. He told Meaney about these facilities and made arrangements to drive by the barber's home and take him there the following Saturday. Meaney was so inspired by having access to the tools of the etching trade that he went to the library and read for months. "I got so I could print fairly well," he reported. There were weekends when Ding was busy elsewhere or traveling, which interrupted Meaney's progress. Finally, Darling gave him a set of keys to the house and told him to use the place whenever he was free and wanted to do some printing. Darling would occasionally stop by the Meaney house after a morning at the Peony Farm. "He would have a beer and a sandwich and we'd talk. He knew more about everything than anybody I ever knew," Meaney recounted.

Ding was Meaney's mentor and, through Darling, he became so proficient that he did printing for such artists as Sue Fuller, Samuel Chamberlain, and Mauricio Lasansky. The Dutch artist Cornelius A. Bartels came to the United States and was a visiting artist at Central College in Pella, Iowa. While in Iowa, he became acquainted with Darling and helped the cartoonist add detail to his etchings. Robert Colflesh— Des Moines attorney, World War I hero, and political ally of Darling's—also became a skilled printer and joined Meaney and Ding on Saturday mornings at the Peony Farm, where they became something of a team. Meaney and Colflesh later printed many of Darling's etchings.

Although Darling was more than twenty years older than Robert Colflesh, the twosome shared deep personal and political convictions. They were self-sufficient men, who by their own wit and courage had become professionally and financially successful. They shared serious misgivings about the New Deal and the detrimental effects they believed it was likely to have on individual initiative. They arrived at similar philosophical destinations, though by dramatically different routes.

Whereas young Darling had grown up in the company of his parents, Colflesh, from the age of five, lived in the Odd Fellows Home in Mason City, Iowa. He had been born in Des Moines early in 1900. When his mother died in 1905, his father placed Robert and his four siblings in the home. Seven years later when his father remarried, Bob rejoined him in Des Moines. The youth attended West High School, where he played football and was a member of the state championship basketball team.[6]

As a boy, Colflesh spent much of his time as Ding had. Young Bob tramped, hunted, and trapped in the countryside. At ten he and a friend were running five miles of trapline, skinning skunks, muskrats, and other fur-bearing game and selling the pelts to a Mason City furrier.[7]

When he was seventeen, he wanted to enlist in the Aviation Corps but was too young. He somehow convinced authorities in Denver that he was of age and became part of Company M of the Seventh Infantry. He and his freshly trained company were almost immediately sent to the front. Colflesh had never fired his rifle until his first engagement at Ciercy, France, where he was wounded in the throat.

Just before dawn July 15, 1917, Colflesh's platoon was caught in the open at Fossey, near Chateau Thierry. Shrapnel cut the men about him to pieces. Colflesh was hit in the shoulder, in the left knee, and the left ankle. He lay in the mud and blood, wedged between one dead man and one dying man for fourteen hours. Litter bearers finally retrieved him from the field. Battlefield surgeons unceremoniously sawed off his leg just below the hip. He was put in a tent with other young men who were not expected to survive. Colflesh was later treated in a series of eight hospitals. Shortly after he was released from Walter Reed Army Hospital in Washington, D.C., he found a job as a janitor, but he was not yet strong enough to keep it. Congressman Cassius Dowell of Iowa helped him enter George Washington University. The young veteran took time out from his college studies to learn shorthand and worked the rest of his way through school as a clerk on one of McDowell's committees. He completed eight years of college work in seven years, earning a bachelor's degree and a law degree.[8]

A decade after he was wounded in France, Colflesh returned to Des Moines to practice law. A recipient of the Distinguished Service Medal and the French *Croix de guerre,* he handled hundreds of compensation cases for disabled veterans without charge for his services. He later served as district census supervisor, U.S. district attorney and Iowa American Legion commander. In 1934 he was a candidate for the Republican nomination for governor of Iowa. In the mid-1940s he contracted tuberculosis, a result of a 1917 mustard gas attack in France, and spent a year in Veteran's Hospital in Des Moines, where he continued to do legal work. (He had been to the office every day of the previous week when he died Monday, April 17, 1967.)[9]

Colflesh shared Darling's appreciation for nature and especially Ding's enthusiasm for hunting and fishing. He and Darling became acquainted in the 1930s, apparently first meeting in the hallways and elevators of the Register and Tribune Building, where the attorney's firm—Parrish, Guthrie, Colflesh & O'Brien—was officed.

When Colflesh and Darling went to the Peony Farm, the attorney's wife Chloris would fill a thermos of hot soup, and send along crackers or sandwiches. "Those were some of Bob's happiest times," she said. One winter, when Jay was in Florida, Colflesh and Meaney decided to send their teacher a portfolio of the Peony Farm workshop. Chloris went out with her camera, she remembered, "and we took pictures of all aspects of making an etching, and even the corner drain which in a pinch served as the 'rest room.'" She made prints of the photographs in her darkroom and Colflesh and Meaney put them, along with other objects, in a huge scrapbook, which they sent express to Florida as a Christmas card for Ding. In return, Ding sent Chloris a sketch of himself ushering her into the Darling "Academy" as an honorary life member.

When Darling returned to the Register and Tribune Building, from Washinton, D.C., the huge wooden desk from which he had directed the resurrection of the Biological Survey was sent to him, as was customary. Darling never used the desk. Instead, he sat the diminutive Merle Strasser behind it, and she used the mammoth piece of furniture exclusively for years. Darling was accustomed to doing all his sit-down work at his drawing board and, even though Merle always thought he should have a desk, he continued to write, dictate, sketch, and confer at his drawing table.

Darling's spelling had improved little since his days as an undergraduate at Beloit College. Even so, by the 1930s he had become a prolific and effective writer in conservation circles. The accuracy of his business correspondence was a credit to Merle Strasser, who took pains to correct his handwritten notes as necessary. She once transcribed a letter in which he referred to guerilla warfare as "gorilla" warfare. "I still laugh every time I think about that," Straser grinned. Then she laughed.

Darling's secretary typed a thirty-year, nonstop stream of correspondence for him, as he left her notes and sent her Dictabelts from wherever he happened to be. His correspondence often bore a Strasser note including the dictation date and place along with the transcription date and Des Moines address.[10] Darling would occasionally ask Strasser as soon as she had transcribed a letter, "How does it read?" She would reply, "I can't tell until I read it." Darling never seemed able to understand why Strasser could not read a letter at the same time she transcribed it—while watching the spelling, listening for his pronunciation, and correcting the grammar on the Dictabelt.

Darling was self-reliant and rugged. He refused to be discouraged. It was not his nature to complain, according to Strasser, "and he didn't appreciate complaints from others." Yet Darling, the fiscal conservative, was a generous man. "He would hand over money freely to anyone who needed it," said Strasser. His generosity was often felt by his secretary. She once took a cake to the office in honor of Darling's birthday. Darling later learned from a third party that Merle had no electric mixer. After lunch "he came in and put a Mixmaster on my desk," Strasser recalled. He said simply, "This is for you." Then he went about his business. He exercised restraint, but he "was very emotional. He wanted to help. But if someone was trying to 'work' him, he could see through" the ploy and the offending party would be out on his ear.

The energetic, imaginative, affable would-be doctor, however, "had no truck with people who were sick," Strasser noted. He was uneasy in the presence of serious illness and "he was not one to call on the sick." It was almost as if he thought if he ignored illness it would go away.

Merle Strasser and Harold (Tom) Carlisle were members of the Darling team, and their responsibilities as aides to the effervescent Ding were as varied as the cartoonist's interests. Darling one day bounded into the office with a large, live turtle beneath his arm. He handed the creature to Merle and asked, "Do you want to chloroform this fellow?" The shocked secretary knew Darling was not asking; he was telling her. Darling needed the turtle to make a mold for a metal casting. He had designed a creative sundial to be placed in Greenwood Park in Des Moines; a sundial borne on the back of a turtle. The team of Strasser and Carlisle administered the lethal dose and Darling immortalized the creature in bronze.[11]

Carlisle was proof of Darling's theory that any one with intelligence and diligence could become a working cartoonist. In 1926 Ding had bet *Register* editor Harvey Ingham that a frustrated University of Iowa student of taxidermy could be trained, under the cartoonist's tutelage, to become a competent artist. Darling needed a skilled technician who could occasionally do the finishing work on his cartoons and could draw cartoons of his own for the *Register* when Darling was absent on his rounds of speaking engagements or while he was vacationing or recuperating. The experiment worked out so well that Carlisle succeeded Darling as the *Register*'s cartoonist when Ding retired in 1949.[12] Strasser recalled Carlisle as a "pretty good" cartoonist but one who lacked Darling's diligence and drive for perfection. "He was a good artist," but he didn't have Ding's "sense of humor, kindliness, education, or broad scope of interests," she said.

Darling had many imitators, including Carlisle, Kenneth MacDonald emphasized, "but none was able to instill Darling's quality of motion in a cartoon—the illusion that, if you glanced away, the cartoon would move by the time you looked at it again." Tom Carlisle's career with Darling required that he imitate the Ding style of cartooning. Their work seemed identical, according to MacDonald, "but to the close observer they weren't at all. Carlisle was never able to instill that sense of tension in his work, that vibrancy, that motion."

The experiment that had been so successful in the short run failed over the long pull. Long after Ding retired, Carlisle left newspapering and cartooning, even though he was only about fifty years old. MacDonald tried to dissuade Darling's protégé but became convinced that

Carlisle had confronted an important revelation; he had been trained to work with a style that was not his own, and he could no longer accommodate the schism.

Not only was Ding a nationally renowned cartoonist but his stint in Washington had made him a nationally renowned conservationist as well. His moves were carefully watched and amply commented upon. His fame was attested to by a letter addressed only with a drawing of a bell being struck by a hammer and "Des Moines, Iowa." It was delivered to Ding two days after being postmarked in Philadelphia![13]

He had made some commitments to himself and to conservation when he left the Biological Survey. He was inspired by the notion of a federation of conservation organizations that would centralize the educational process and, equally important, marshal the political power required to influence legislation. In a rare defense of Franklin Roosevelt, Darling wrote, "It has been a discovery painful to me that many of our public officials, including the President and Ickes, have a much more favorable attitude toward conservation or wild life than they are ever able to exercise because of the fact that exploiters of natural resources are always organized and the conservationists never."[14]

Darling noted that "eleven million horses running wild couldn't pull a rubber-tired baby buggy to town unless there was a harness to hook them to the load." Likewise, he insisted, eleven million sportsmen and 36,000 scattered sports groups "should have some kind of harness to band them together to exert a united influence for the good of wild life."[15]

Jay Darling had hardly warmed the chair in his *Register* studio, after his stint as Survey Chief, when he involved himself in efforts that led to the creation of the National Wildlife Federation.

# 15

## "The Baby Is Yours"

So MANY SEEDS were planted at the New York City dinner meeting with the arms and ammunition manufacturers in April 1935 that sprouts were emerging from the soil soon after Jay Darling again took up his cartoonist's brush at the *Register*. With the pledge of assistance obtained at that meeting, Darling and Senator Fred Walcott of Connecticut had called another, even larger conference for July 16, 1935. Its result was the creation of the American Wildlife Institute, whose role was to carry on the expanded public activities of the American Game Association, which subsequently became inactive. The new American Wildlife Institute was officially established August 20, 1935, and Thomas H. Beck was elected its first president.[1]

Before Darling left the Biological Survey, he had also urged President Franklin Roosevelt to issue a call for a huge North American Wildlife Conference, as a successor to the American Game Conference that had advised presidents for years. Roosevelt agreed and the conference was scheduled in Washington, D.C., February 3 through 7, 1936. It was a rousing success and was attended by representatives and participants from throughout the United States and Canada. F. A. Silcox, Darling's Washington colleague and chairman of the conference, closed the meeting with the words, "Jay Darling, God bless you, the baby is yours now." The nearly 1,500 conservationists who had adopted a constitution for the General Wildlife Federation had also elected Ding Darling its first president.[2]

Even before the February meeting Darling had laid plans for creation of the General Wildlife Federation and had already urged that it be kept free of the American Wildlife Institute and any suggestion that industrial interests were determining the organization's policies. Even though manufacturers had provided the opportunity to create an effective wildlife organization, Darling was conscious of public opinion and wary of unnecessary misunderstandings that could blunt the Federation's effectiveness. As chief of the Biological Survey he had steadfastly defended the role of the industrial interests and attested to their hands-off posture, but he wanted to avoid such brushfires in the future. Darling's position seemed especially wise that fall when Walter Chrysler, original chairman of the American Wildlife Institute's board of directors, was arrested on Chesapeake Bay. Chrysler was charged with illegally baiting blinds over which he was shooting and with exceeding the bag limit. He resigned as chairman but remained as an Institute trustee.[3]

Public relations problems associated with the similar names and purposes of two distinct organizations continued to fall on Darling's desk. In making his explanations, he was as careful to defend the sportsmen as to explain the purposes of his Federation. An outraged woman, vice-president of the Anti-Steel Trap League, wrote Darling, "By mistake, thinking it was your organization, I joined the American Wild Life Institute, and was later disgusted to find it just a vehicle for the hunters and munition-makers, or chiefly that." Darling replied that the munition makers wanted something to do and that they had been so callous as to give $170,000 to a board of wildlife experts to spend as they saw fit for the cause of wildlife conservation.[4]

Darling's plan was based largely on an organization of conservation interests spearheaded in Indiana by the young, energetic director of education in that state's Department of Conservation, C. R. (Pink) Gutermuth. Darling first met Gutermuth in the fall of 1934 at a conservation meeting in Chicago. Gutermuth had already spent a year shaping up a statewide organization of sportsmen's clubs. By January 1934 he had 508 active conservation clubs in Indiana, all local and individual in every respect. Each of the clubs named delegates to their respective county conservation councils. One person was elected from each county to attend meetings at the conservation district level and to elect a district representative to the state Conservation Committee. Ira Gabrielson and Pink Gutermuth later became president and vice-president, respectively, of the Wildlife Management Institute, successor to the American Wildlife Institute.[5]

The plan for a General Wildlife Federation, announced by Darling before the first North American Wildlife Conference, was similar to the Indiana system and one established in New York about two years earlier. One writer said, "Such was the magic of Darling's appeal that within five months after he announced the plan, twenty-five states reported the recent mobilization of their conservation clubs under a central state federation; fifteen already had existing state sportsmen's organizations; and four others were on their way toward federation." [6]

While organizing and overseeing the efforts of the General Wild-life Federation, Darling also kept his hand in other conservation activi-ties at the state and national levels. With his prodigious letter writing, he rained a continuing barrage of comments and observations on the conservation scene.

He noted that his successor at the Biological Survey, Ira Gabrielson, "is performing wonderfully. He will make a much better chief of the Biological Survey than I ever could be and I am very happy over the results." Darling lent his hand to getting a Civilian Conserva-tion Corps (CCC) camp reopened at Milford, Iowa, for the benefit of the Okoboji Lakeside Laboratory operated by the University of Iowa in Iowa City. The CCC camp had operated for two years but had been closed in October 1935. E. A. Gilmore, president of the university, re-quested Darling's help in getting the priority rating of the camp moved up on the Department of Agriculture's waiting list. The reopening of the CCC camp was approved in March 1936 after Darling appealed to Henry Wallace and others to reactivate it. The Lakeside Laboratory had been established in 1909 on West Okoboji Lake to provide a place where the rich fauna and flora of the northern Iowa lake and prairie regions could be studied and conserved. The site was deeded to the state in May 1936. [7]

Darling's health failed again in the spring of 1936, forcing him to rest from about mid-March until May. The effects of his illness were such that his physician ordered Darling to maintain a slower pace, to cut down his traveling, and to quit making speeches. By early 1937 it was rumored that Darling would soon resign his post as president of the General Wildlife Federation, and some of his disciples expressed their concern for its future if that should happen. An interested observer wrote Darling, "Neither the Federation nor other interests are yet strong enough to stand alone or to make progress without your leader-ship." Darling's secretary responded, "I am unaware of his proposed resignation as President of the General Wildlife Federation but presume his doctor has advised it. I know he forbid [sic] Mr. Darling to make speeches, especially those that involved traveling some dis-tance . . . as that seems to fatigue him more than anything else he does." [8]

Darling took advantage even of his poor health to immerse himself in a new experience. He had spent several winters in Florida as a means of escaping the deadly effects of Iowa's winter weather on his respiratory ailments. In the winter of 1936 and 1937, however, he decided to make the trip to Florida in one of those ''new-fangled'' house trailers:

The immediate excuse for taking such a journey was with us, twofold; first: a case of chronic bronchitis which wasn't going to be helped materially by the next three months of coal smoke, ice, sleet and aspirin tablets; second: a somewhat fantastic idea that if the whole pattern of American life was to be completely altered by this new device of perambulating penthouses, it was up to us to acquire some advance information about it.

The trip was chronicled by Darling in a small, entertaining book, *The Cruise of the Bouncing Betsy*, spiced with Ding's humorous drawings. Darling was sold on the comfort and convenience of trailers for those who had some pioneer spirit left in their veins, but he discovered he was years ahead of adequate trailer parks.[9]

The journey ended when the Darling's trailer and automobile made the ferryboat trip from Punta Rassa, twenty miles from Fort Myers, Florida, to Sanibel Island off Florida's west coast. Finally, Darling drove over a one-way bridge and down a winding shell road to 'Tween Waters Inn on Captiva Island, which was probably his favorite spot on earth. The island was a largely undiscovered paradise when he first visited the 'Tween Waters in 1935 on brother Frank's recommendation, and Darling wished passionately that somehow it could be rescued from the disastrous effects of the ''progress'' he foresaw.[10]

At that time, a boat was the only means of communication between the mainland and the picturesque island in the Gulf of Mexico.

There are no telephones and the mail boat comes once a day when it runs and drops us the four-day-old newspapers from New York and home. Who cares? This is no ordinary tourist resort. There is not a dance floor on the island and no raucous jazz music drowns out the soft singing of the breezes through the cocoanut palms at night. . . . 'Tween Waters is owned and managed by a Prince and his Charming Consort who has a rare genius for fine flavoring of hospitality as well as food. We do not dress for dinner and the fishing is marvelous. I hope no one else ever finds it—least of all the trailer multitude. To invade this sanctuary of nature with a trailer colony would be as out of place as taking a hurdy-gurdy to prayer meeting.[11]

Ding's travelogue, although it was entertaining and informative, was not a best seller. Six years after it appeared, the publisher informed Darling that even if the entire remaining stock of the book were to be sold royalties would not repay the company the advance paid to author Darling.[12]

Darling maintained his steady flow of correspondence from Captiva each winter, even though there were considerable delays in his responses. Merle Strasser sent the important incoming mail to Florida where it was put aboard the mail boat; Darling sent his answers via the same slow craft. He also sent his cartoons for the *Register* and the *Herald Tribune* syndicate from Captiva. When he returned to Des Moines, however, his correspondence picked up its normal speed. He sent the president of the University of Iowa a curt telegram expressive of the Darling impatience: "Meet me tomorrow at 12 noon my office Register and Tribune Building. Do not disappoint me. J. N. Darling." The reason for the urgency was not revealed in the telegram nor in President Gilmore's negative reply. Darling hammered away for a CCC camp at Hampton, Iowa, with letters to Iowa senators Guy M. Gillette and Clyde Herring and to Conrad Wirth, director of the National Park Service. He sent a telegram to Alf Landon asking the unsuccessful 1938 Republican presidential nominee to straighten out Kansas Congressman Clifford R. Hope, who seemed bent on emasculating "the best conservation bill in the history of wild life efforts." He also wrote Hope, after the bill was pocket vetoed by Franklin Roosevelt, and apologized for the "heat we turned on you. . . ."[13]

In the spring of 1937 he was in the forefront at the Saint Louis, Missouri, meeting of the General Wildlife Federation, which was referred to there as the American Wildlife Federation, and was reelected president of the organization. (The name of the group was officially changed to National Wildlife Federation in 1938.) He completely dominated the gathering, according to one observer:

But what a thing it is to see a man like this Darling emerge from the ruck and rumpus of controversy, jealousy and pessimism, and throw his efforts into a thing that everyone knows is necessary, that some one has got to do. . . .

Nor did he know, when he accepted the job as chief of the bureau of biological survey, that he would be shoved around by the various governmental bureaus as he was. He quit the job and set his heart upon uniting the conflicting groups. He had made a good start. Maybe that is all it will ever be—a good start. It's been undertaken six times in the past and always has failed. It may fail again.

The writer described Darling as a man who "can orate and he can snarl. He can bare his teeth and rear back on his haunches and let drive at the malcontents and racketeers."[14]

In accepting the presidency of the organization, Darling was quoted as saying, "I will do my best but I am in the used car class." He was cast as a martyr to conservation: "The man is tired. In the last two years he has suffered two nervous breakdowns. It is not generally known and his friends urge silence on this score. Why? Is it because they are afraid that the physical frailty of their inspired leader may do harm to the national movement? Perhaps that is so. It is too bad.'' (Darling's friends, co-workers, family members, and acquaintances discounted the occasional rumors of Darling's "nervous breakdowns.'' His physical condition was fragile, and he did little to guard his health, they acknowledged. As he grew older, he suffered more and more frequent physical debilitation. His correspondence, however, reveals no disorganization of his thoughts, no serious inconsistencies, and no letup in 1935 and 1936.)[15]

People who had known him as a cartoonist were amazed at the depth of his feelings on conservation. And were amazed, too, at the deftness with which he engineered the organization of the permanent set up. People came and went on the platform in the big gold room of the Jefferson Hotel. When Darling got tired of talking, he'd consult a list of names and summon some expert to take over for a while. Then he would return to the fray, bristling, humorous, kindly beligerant [sic], changing his voice and manner as the situation demanded.[16]

One accomplishment of the 1937 Saint Louis conference was the establishment of National Wildlife Week, with the Federation taking the lead in its organization and promotion.[17]

The third meeting of the Federation took place in Baltimore in the spring of 1938. Darling was again elected the organization's president, and its name was officially made the National Wildlife Federation. Darling, insistent upon making the Federation financially self-supporting without levying taxes on the local conservation clubs, put his artistic talents to work. He designed the first in an annual series of wildlife stamps to be offered for sale to the general public, and this system of independent financing is still used successfully by the Federation.[18]

While Darling was making progress as no one else had in consolidating conservation interests in the United States, the most tragic incident of his life loomed just beyond the horizon.

Darling's son, affectionately known as "Dr. John," had interned at Cincinnati Hospital and was a fellow at the Mayo Clinic in Rochester, Minnesota, where he had earned respect as a bright, accomplished young physician. He was the very Doctor Darling that Ding had hoped to be, and he thrived in the atmosphere of the medical profession.

January 13, 1939, Dr. John left Rochester for Des Moines, driving an automobile belonging to Mrs. Arthur Neumann. Mrs. Neumann was in Rochester recuperating from surgery, and her husband had remained there with her. The Darlings and the Neumanns were friends, and John was apparently returning the car to Des Moines at Mr. Neumann's request. It grew dark and threatening as John pushed southward. By the time he neared New Hampton, Iowa, a light snow was falling. The temperature had been mild during the day and the falling snow melted as it struck the roadway. The auto suddenly slipped from young Darling's control, skidded on the wet road, and smashed into a highway bridge approximately one mile north of New Hampton. The twenty-nine-year-old intern was thrown from the car. His left arm was broken and he suffered a basal skull fracture. He was hospitalized at Saint Joseph's Hospital in New Hampton in critical condition, and was later transferred to the Mayo Clinic.

Jay and Penny were inaccessible on their Captiva Island retreat, where they had returned from several weeks in Guatemala. It fell to Mary Darling Koss and her husband Richard, living in Des Moines, to rush to Dr. John's side, accompanied by Dr. Lester D. Powell, a Des Moines surgeon and friend of the Darling family.[19]

When word of the accident reached the Darlings, they immediately traveled to Rochester to join the Kosses at the bedside of their seriously injured son. Ten days following the accident Merle Strasser wrote a Darling correspondent, "Dr. John is now practically out of danger and the period of suspense is over, but Mr. Darling expects to remain with him for several weeks."[20] Darling's secretary could not know that the period of suspense was not over for John Darling and his family; that the auto accident had done brain damage that would truncate the young doctor's brilliant career and make him an invalid years after the auto wreck.

Ding attended the fourth meeting of the Federation in Detroit in February 1939, and there handed in his resignation. *Time* magazine described him as "disgruntled," even though under his leadership the Federation had seen the Pittman-Robertson Act set aside the ten percent excise tax on sporting arms and ammunition for wildlife propagation and research. (The legislation was actually passed during Darling's term as chief of the Biological Survey.) The magazine saluted Ding's efforts: "Hunters and animal-lovers, unified at last, have pushed through many a national and state fish-and-game law." Whether the *Time* writer was aware of it or not, Darling was preoccupied with the health of his son. Immediately following the Detroit meeting, the past-president of the National Wildlife Federation went to Florida, his secretary wrote, "for a little rest and vacation, taking with him his son who is recuperating from a serious automobile injury."[21]

Darling was succeeded as president of the National Wildlife Federation by David A. Aylward of Boston. Aylward "carried the tough responsibility of guiding the group through the perilous waters of financial distress and heartbreaking decisions,'' according to Carl Shoemaker.[22] Aylward was also at the controls during the time that Darling lost most of his confidence in the Federation and criticized its lack of direction and loyalty to its original objectives.

World War II was about to entangle the United States and upset the lives and objectives of all its citizens and its institutions, including conservationists, the National Wildlife Federation, and cartoonists.

# 16

## Clouds of War

JAY DARLING had a voracious reading appetite that required the digestion of at least six newspapers each day; piles of correspondence with influential friends; and countless books, magazines, journals, brochures, and pamphlets. His livelihood depended upon timely interpretation of current events; and his eager mind continually processed the data received from his reading, travel, and discussion. As one result, he became convinced that the United States would be drawn into war. In June 1940 he joined five other prominent Iowans who publicly urged President Franklin Roosevelt to come to the aid of the Allies, even if it meant a formal declaration of war.[1]

As American involvement in the war inexorably approached, Darling could see that conservation was going to suffer. It would be shoved into the background in favor of priorities related to more immediate challenges to human survival.

The coming of the conflict, coupled with another episode of overexertion, also disrupted what might have been a successful experiment in combining conservation and the respected Chautauqua summer program in New York. Late in 1940 Darling had busied himself with plans for a program of study in conservation to be presented at Chautauqua the following summer. He was obliged, however, to notify Dr. Arthur E. Bestor, Chautauqua Institution president, that he could not complete the assignment. Darling had suffered a serious recurrence of stomach ulcers and had turned the planning of the program over to Aldo Leopold. Darling had lived with stomach ulcers more or less intimately for years, he revealed, but generally kept them under control with periods of rest and rigid dieting. The Chautauqua conservation program, originally planned for 1940, was canceled that year because speakers were unavailable. The 1941 program was ultimately canceled as well due to international events.[2]

Darling's physician again clamped down on the overactive, sixty-five-year-old cartoonist. Ding was ordered to restrain himself for nearly a year and to avoid making speeches, but he did not follow his doctor's orders religiously: "I have tried it [public speaking] just once since my managed program went into effect a year ago and did a little piece of evangelizing at an Izaak Walton League banquet and it took me two weeks to recover from the effects. The audience probably got over it by the next morning, as most conservation audiences do."

Darling worked on a reduced schedule, however. He claimed that for more than a year he worked half time on his newspaper job. He admitted that he was "doing an occasional cartoon, but that is stretching the doctors' orders a little." Those doctors had told Darling that he had heart problems, but the irrepressible Ding refused to believe it. He wrote Gardner Cowles, "The next thing I knew something hit me that the local doctors called a heart attack. Frankly I didn't believe them . . . my own diagnosis . . . allows me many of my accustomed privileges which the doctors denied."[3]

Darling's physical condition drained him of energy and initiative and left him feeling heavily burdened. His load became crushing in late 1942 when his associate at the *Herald Tribune,* a cartoonist who would "deliver cartoons whenever I didn't come through, had a crackup this spring and has been off watch ever since and probably isn't going to ever get back to work."[4]

Ding, who labored intensively until the job was done, had difficulty doing his work, but he also had trouble grasping the concept of the forty-hour week. His cartoons, especially during World War II, reflected his disdain for such an arbitrary measurement of accomplishment. He insisted that the forty-hour week was conceived to spread what little work there was around to more workers and thus ease the misery of a great many unemployed. To Darling, however, the forty-hour week was still the symbol of a short work week. He further asserted that the forty-hour week was not in the national military interest: "If we don't want to lose this war we had better leave such domestic adjustments as wages and hours until after this war is won."[5]

The conservative cartoonist was predictably opposed to Social Security, although because of his wealth he sometimes hesitated to publicly criticize the system. In a letter to Elmo Roper, however, Darling wrote, "I wish I had a good idea for a cartoon every day too, and it is just as practical for me to expect the government to guarantee that I shall have one as it is for any known form of social organization to guarantee social security." Social Security or not, Darling was thinking seriously of retirement. He was sixty-five, his health was more delicate than ever, his work load was heavy, and he was convinced his style was becoming dated and passé. He wrote to a fellow cartoonist, "You must know that as cartoons go my style of drawing and my conception of cartoon composition is on the way out. . . . The Duffy-Fitzpatrick school is very much in the ascendancy."[6]

Darling was also disgruntled with the editorial direction of the *Register* and with some of the opinions of W. W. (Bill) Waymack, who disagreed repeatedly with the content of Ding's cartoons. Darling did not mince words:

As to the Des Moines *Register* and Minneapolis papers, I have long since ceased to struggle over their editorial skimmed milk. Vigor of editorial convictions and forceful expression throughout the newspaper world seems to have kept company with the general deterioration of American morale, and to have slipped down to the same level of softness which led us into a war with neither equipment nor determination.[7]

The core of Darling's disagreement with the *Register*'s cartoon policies was addressed later by him in a letter to editor Waymack:

In reference to this cartoon business, I am somewhat embarrassed by your solicitude over what might roughly be called my "personal feelings." I really haven't any. If maintaining the cartoon as a standard feature is not to the best interests of the newspaper, it is of no interest to me. . . .

I have always been afraid that sometime I would find myself working for a newspaper that had a candidate who had to be supported because of some personal relationship rather than for the principles he represented. That has spoiled more newspapers than any other one thing I know of. We have done it twice and each time lost ground. Once was when Harvey Ingham got bitten by the senatorial bug and the other was when we went all-out for Dickinson.

Darling then revealed a shocking intention:

I am not opposed to Willkie. In fact I would just as soon he would be the martyr in this coming election as anyone on the list, but I never have had a sacred cow and never hope to have one. If we cannot say (with a chuckle) what everyone else is thinking about Willkie, or anyone else, then it is no fun, and that at least is a large part of cartooning.

As a matter of fact, I don't find any keen, thoughtful leadership anywhere in the Republican Party. It is a mess and ought not to be flattered by sober support. The best I can do under the circumstances, so far as the Party is concerned, is to do a not too caustic lot of ribbing and perhaps a reluctant vote on the Republican side just to keep the two-party system alive—although if no-one is looking I will probably vote for Roosevelt.[8]

Waymack wrote the *Register*'s editorials. Gardner Cowles was still at the *Register* and *Tribune,* but his health had declined; he had lost his sight, and he could hear only well enough to be read to. His son, Gardner (Mike) Cowles, Jr., had assumed management of the papers. Darling, who was accustomed to answering only to Gardy, was less than happy with the arrangement. He referred to Mike Cowles and his brother John as the "kids," although there was respect and affection between Darling and the two Cowles brothers. When John moved to Minneapolis to head up the Cowles newspaper enterprise in that city, he lamented his lost opportunity to swap ideas with Darling. Both brothers sought and received Darling's advice on business and personal decisions. Things, however, just were not the same. Darling announced his plans to quit. In a letter to Grantland Rice, Darling reported, "I quit working about six weeks ago, I mean retired, signed off, and other words to that effect, and have been enjoying the luxury of waking up in the morning without any deadline hanging over my head. It's great stuff, but I don't know whether I am going to get away with it permanently or not. At least I am enjoying to the limit the thought that I don't have to do it anymore."[9]

Ding was through and only waiting for the irritating details of his newspaper association to be cleared up. Gardner Cowles was distressed at the prospect. Largely cut off from the operation of the papers by his health, the elder Cowles, in dejection, once asked his secretary, "Agnes, why on earth did I lose my eyes? I have been a careful man. I haven't done any dissipating, and here I have lost my eyes." Cowles did not want to be cut off from Ding, the bright, brash young cartoonist he had lured from the Sioux City *Journal* nearly forty years earlier. He wrote, "Personally, I should very much regret having you sever your connection with the *Register* and *Tribune.* I think you know full well that I have been your warm friend and sincere admirer for about thirty-five years or more. I hope at least that you will continue with the *Register* and *Tribune* during the brief period that I will probably be able to continue here. If there is any irritation I believe it can be adjusted."[10]

The relationship between Cowles and Darling, particularly as they became the old guard at the *Register* and *Tribune,* was as intimate as possible for two aging and reserved sons of clergymen. Cowles, beset by physical afflictions, had earlier written Darling simply "to tell you how I value your friendship and how much satisfaction I have derived from our close relationship for many, many years." In spite of his occasional disagreements with the crusty publisher, Darling in response recollected their partnership in roseate terms, "The Lord knows I've been highly paid for my funny pictures but you were always one to squeeze on the contract and then pay double before the contract was half over. And the best reward of all has been the greatest personal satisfaction which can come to anyone: a lifetime of work under a perfect boss."[11]

Just as Gardner Cowles was imploring Ding to remain at his drawing board, the Pulitzer Prize committee named Ding winner of his second Pulitzer award, this time for a 1942 cartoon depicting Washington, D.C., inundated in a sea of paperwork. Darling was flabbergasted. "I wonder how they pick 'em," he wrote. "I even had to look in the files to recall that wholly insignificant and uninspiring cartoon on which they pinned the blue ribbon." Robert L. Ripley was one of the hundreds of fans who wrote to congratulate him on his second Pulitzer Prize. Ripley wrote, "Friend Ding you have always been a prize winner to me since that day I met you in the Singer Building thirty years ago, believe it or not."[12] Darling wrote in response to another congratulatory letter:

That Pulitzer Award came darned near being a posthumous award, figuratively speaking, for about six weeks ago I decided that I had drawn cartoons long enough and sent in my resignation. I didn't seem to be getting anywhere and I'd been at it for almost forty years and it seemed to me it was time to lay down the shovel and the hoe and hang up the fiddle and the bow like old Uncle Joe, and watch the world go by. I guess I'm not going to be able to get away with it.[13]

On the heels of the Pulitzer Prize, Darling was named the winner of the Roosevelt Medal for his conservation work. A glowing letter from Teddy Roosevelt's partner in conservation, Gifford Pinchot, congratulated Darling for winning the medal. In reply, Darling questioned whether anyone ought to get a medal for conservation. He likened the ritual to giving medals to generals who do nothing but lose battles.[14]

The trend in conservation obviously troubled Ding, and the war itself had brought into sharp focus in his mind the necessity for conservation and the realities facing the world's growing population. He wrote, "Continually increasing populations in the face of perpetually decreasing resources always have and always will generate bloody warfare, in every species from man down to the microbes. There are not exceptions. The present war-torn world is little more than that although you call it Bolshevism, Naziism, Democracy and what-not." It was equally difficult, he wrote, to make Americans see the danger of world food shortages: "I think I know the reason why the American public can't rationalize on the subject of bulging populations and decreasing resources. It seems quite impossible for a man who has never known anything but a full stomach to think in terms of starvation."[15]

By the fall of 1943 Darling was urging Iowa's Senator Guy Gillette to study carefully the impact on soil and water of some postwar projects, particularly dams, already being contemplated by Congress. At the same time, he was pushing Iowa's governor to set up a professional state conservation organization and to rid the Conservation Commission of political overtones.[16]

The war had also brought violence and disruption to Darling's Sanibel paradise, and he felt called upon to pitch in there as well. The Coast Guard had driven the small commercial fishermen off the Gulf waters, ostensibly for their own security, and bombers from the Fort Myers training field were using Upper Captiva for target practice night and day. By May of 1942 Darling had completed his unique Fish House, an elevated retreat off the shore of Captiva where he could raise a counterbalanced drawbridge behind him and work without interruption. Darling had made Captiva a part-time home and, war or not, he was determined to make Sanibel and Captiva islands a federal refuge, maintained by the Fish and Wildlife Service. A proposed sale of island land to a real estate dealer was narrowly squelched when Florida Governor, and later U.S. Senator, Spessard Holland, twice postponed the public sale of the state-owned lands at the request of Darling and others.[17]

Darling, closer to home, also paid the costs of an experimental short-course school in conservation, designed for teachers and cosponsored by the League of Women Voters. Marguerite M. Wells, president of the national League of Women Voters, remembered Darling as a tower of strength to the League in its early days. Darling's opinion of the League had deteriorated. "I have come to the regrettable conclusion," he wrote, "that the League of Women Voters is interested in conservation only when someone will pay their bills or speak to them and furnish all the scenery."[18]

On the national conservation front, soon after his resignation from the National Wildlife Federation's presidency, Darling perceived that organization being transformed from an advocate and coordinator of wildlife programs in general to a pressure group for the Fish and Wildlife Service. He made a distinction: "As organizer of the Wildlife Federation I can say quite honestly and frankly to you that we were not organized to serve as a pressure group for the Biological Survey but to coordinate the thousand and one little groups of conservationists which exist throughout the country but have no unity of purpose or coordination of program."[19]

Darling wrote Charles W. Collier, executive director of a new organization called Friends of the Land, encouraging the formation of the Friends but warning him of the many pitfalls that cut short the progress of the National Wildlife Federation. Darling was disappointed at the lack of cooperation he had seen among the various and diverse conservation groups called together to form the Federation. Each cell within the Federation's body seemed to have a distinct purpose, function, and interest. Self-serving politics and jockeying for favorable position had wrecked the Federation's intentions and made a mockery of the organization's name, Darling believed.[20]

The Federation was having difficulty maintaining an effective staff and, in a related realm, was having little success finding the money it needed to do the job it had set out to do. "I think," wrote Darling, "that I have never suffered so great a disappointment as the failure of that organization to live up to its high potentialities." He wrote to Ira Gabrielson, director of the Fish and Wildlife Service, "If the Federation has to pass out of the picture, as I think it must, this is the most propitious time to hold a very quiet funeral. The complete absorption of the country in the war crisis will make the death of the Federation pass unnoticed."[21]

The war had also disrupted what Darling saw as some positive developments in conservation at the national level of government. To his friend Harold Ickes, secretary of the interior, Darling wrote, "Gabe tells me that in his wildest dreams of effective government service he could not have asked for a better boss than you. The Fish and Wildlife Service is doing splendidly, better than ever before in its history I am sure."[22]

Darling's opinion of Roosevelt, however, had not mellowed:

There is nothing that so completely spoils Roosevelt's day as an intimation that he and his New Deal associates do not, within themselves, contain all the factors necessary to do any and all jobs better than anyone else. I had a ringside seat at his show for two years and witnessed this supreme satisfaction with his own wisdom, and those about him demonstrated many times. . . . Roosevelt hates the New York *Herald Tribune* only a little less intensively than he hates me. . . .[23]

Darling's disappointment in the Federation's activities grew into despondence. He disagreed with staffing decisions that he feared would inject still more politics into the Federation, but at the highest and most influential levels.[24]

While Darling detested the presence of politics in the National Wildlife Federation, he took a renewed personal interest in party politics in his home state. In 1941, with the exponents of the New Deal still in national control, Darling led an Iowa effort to organize the Republican party in hopes of a political comeback. Darling formed what he dubbed the "Committee of Thirty" to make the Iowa Republican party a year-around enterprise for its leaders and followers. The committee members bore some of the best known and most influential names in the state and, in some cases, in the nation. One was Vernon Clark, Darling's friend from his earliest days at the *Register,* who by 1941 was a wealthy owner of Penrod, Judren and Clark Company, which had cabinet wood mills in Kansas City, Cincinnati, Des Moines, and Norfolk, Virgina. F. W. Hubbell and J. W. Hubbell were sons of Des Moines pioneer banker and insurance executive F. M. Hubbell, a man whose estate totaled approximately $100 million in 1977. Others, whose names were associated with nationally known industries headquartered in Iowa, included Fred Maytag, John Rath, and Craig R. Sheaffer.[25]

One major aim of the Committee of Thirty was to make resources available for promoting the Republican party in Iowa. Another, equally important in Ding's view, was to provide better Republican candidates; he hoped to "continue this method of participating in the efforts to provide better government" by means of "this unorganized movement which we have set up." The Committee of Thirty, which eventually became popularly known as the "Ding Committee," was an answer to the threat Darling saw in the alliance of politicians and labor groups: "Business men have a right to have a part in this government as well as organized labor and professional politicians."[26]

Darling's interest in an array of enterprises remained keen and controversial. He continued to serve as a trustee of Grinnell College, urging the school's administration to keep enrollment low while emphasizing quality education, but resigned when policies contrary to his views were instituted. He took an interest in a new torpedo invented by a friend and urged Secretary of the Navy Frank Knox to investigate it. He criticized Harold Stassen for enlisting in the Navy, although he supposed Stassen was thinking of veterans who would be in the voting booths "twenty years from now." He studied carrier pigeons. Forever the frustrated doctor, he offered advice to a Mayo Clinic physician on improving the photographic examination of the alimentary canal. Darling decried the use of DDT for mosquito control and wrote that the number of DDT projects being proposed "scares me almost as much as the uncontrolled use of the atom bomb."[27]

As the war ground on, Darling assessed Joseph Stalin in terms of the "good medicine" he might provide for the United States: "Stalin has emerged as one of the really great men of this generation. He is very able and very cruel and may be expected to interpret international policies on a wholly realistic basis for the single benefit of Russia. I feel certain that there is going to be a terrific collision when Stalin realism meets the American Santa Claus in the post-war negotiations for World Peace.''[28]

The world would soon see the results of postwar negotiations. The day after Japan accepted the Potsdam terms, Darling remained in his office and watched the human eruption with detached cynicism:

The streets are full of tooting automobiles and dancing maniacs. As I look at them from my window, milling around in the street below throwing paper and batting each other over the head with tin horns or whatever happens to be handy, I have a sneaking feeling that they are more excited about the prospect of unlimited gasoline than they are about the more serious aspects of the turn of events.[29]

Time had been lost for conservation efforts, and resources had been recklessly expended in bringing the war to a victorious conclusion. The attention of the American people had understandably been fixed on rifles, not refuges. It was time to get serious about conservation again. It was time to illustrate the relationship between healthy wildlife and healthy humankind. It was time to get busy teaching Iowans and Americans an even larger lesson—''how to make a human refuge on all the land in this continent rather than establish isolated areas for the salvation of bird species.''[30] It was a difficult assignment, and it was going to be a tough fight.

# 17

## Downs and Ups

SHORTLY AFTER the end of World War II, Darling nearly lost his patience with the National Wildlife Federation. He wrote its president, David A. Aylward, "I don't see how the Federation can go on justifying its existence just by selling stamps and having a few executives housed in their Home Office and not making a campaign among the states which might organize and contribute to a national organization."[1]

Darling was equally upset by what he believed to be misrepresentations by Ducks Unlimited of the numbers of waterfowl hatched during the 1946 season. Ding wrote the organization's chief naturalist in Canada to say he had "reservations about your publicity and overstatement of duck populations. . . ."[2] Darling had never seen eye to eye with Ducks Unlimited which, in his view, represented the interests of private shooting preserves and greedy hunters and consistently put the best face possible on bird populations. In 1947 he wrote:

A prophet, they say, is not without honor save in his own country and Iowa seems to have lived up to that old saying for we have no Federation and no representation of the National Wildlife Federation within the State. . . . I dread to mention the hold which Ducks Unlimited got upon the imagination of the folks out here, collecting tremendous amounts of money which, as you know, have resulted in practically no good to the great cause.[3]

Ding was ambivalent toward the Federation. He harbored only a faint hope that the organization would become the force he wanted it to be. He also took its failures personally and was as likely to wish the enterprise dead or to dissociate himself from its activities altogether. "I would like to continue to be helpful but in my present state of mind I know I cannot be that," he informed Aylward, "and would like to drop out of the picture entirely." Three days later, Darling wrote an unsolicited letter suggesting that the Federation move its headquarters to Roosevelt House in New York City. The following month he wrote, "I have seen too many old men who have outlived their usefulness sticking around and succeeding only in getting in other people's way to wish to do likewise myself."[4]

Within days of dictating that observation, Darling wrote of his chagrin at being identified with the Federation at all: "This may sound violent and unjustified to you but by God I'm not going to let my name be used to gyp the public out of several thousand dollars a year to support an organization which has, in my judgment, no justification whatsoever." Soon after, his observations were mellower: "The success of the Federation has never ceased to be to me of more importance than anything else in my life. . . . Having been a party in its conception, no father has been more anxious than I or spent more sleepless nights worrying about it."[5]

Even though Darling believed the National Wildlife Federation failed to unite the nation's conservationists, he fanned the hopes of others. Nash Buckingham had expressed to Darling his dissatisfaction with the Federation's news service, headed by Carl Shoemaker. Buckingham, a member of the Outdoor Writers Association, indicated an interest in organizing a news service analogous to a newspaper or wire service bureau, which would operate out of Washington, D.C. Darling wanted to take up the idea with the publishers of the Des Moines and Minneapolis newspapers—Mike and John Cowles. When the opportunity arose, Darling reported, the brothers "listened attentively to the outline of the service which I presented." He urged Buckingham not to forsake the project because "I believe it is thoroughly practical and one easily sold to the publishers."[6]

Darling was especially interested in seeing a professional news bureau established. Relations of conservationists with the popular press were, in his view, virtually nonexistent. Most newspapers, he suggested, preferred "Hollywood and glamour girls. Don't expect too much from the newspapers, although I say it with shame." Ding had earlier conceived of a conservation clearing house that could publish information for the benefit of all conservation organizations and had gone so far as to construct a prospectus. Its objectives received a warm reception but the means were not endorsed. Nobody had said the clearing house would not help, Darling reported, but "they are afraid of it, thinking that it holds within it a threat to their prestige."[7]

Darling was listening to the same tune but a new stanza. The narrow interests of individual clubs and organizations were about to kill the clearing house idea before it ever took flight. Within a few weeks he gave up. "It already seems to have been time wasted," he lamented. Darling recounted his travails to Buckingham. Everybody thought the information clearing house idea came from "an impractical fuddy-duddy who didn't know anything—meaning one Jay N. Darling. . . . therefore, when no one willingly and enthusiastically backed the project, the Joe Moneybags crowd backed away from it and I don't mind. I'm tired and half sick and if the world wants to go to hell in a basket it's all right with me." Darling's health was again a discouragement to him: "I have almost taken the count with another protracted siege of chronic bronchitis. I am pulling through now but it has been a heavy drain on my energies and I haven't had much interest or time for anything else." [8]

Although Darling was not totally happy with Iowa's conservation program, he believed his state was far ahead of the Fish and Wildlife Service in enhancing the image of conservation. Public relations, he claimed, had always been neglected by Fish and Wildlife but he knew that it could be remedied because the situation had been drastically improved in Iowa. He noted that all Iowa's wardens, who were very poorly paid, were taught annually how to make a speech and what to say and how to approach the public. [9]

Darling was growing restless at what he saw as a tendency on the part of the state's executive branch to meddle in the affairs of the Conservation Commission. Ding held tenaciously to the principle that fees paid by sportsmen and conservationists were not the governor's money nor the people's money, but that those who paid the fees should insist on a free hand in dispensing the funds to technically qualified men in the promotion of their conservation activities. [10]

The cartoonist was troubled as well by the drop in the state's water table and urged that action be taken quickly. He wrote to the editor of the Cedar Rapids *Gazette*, "Our water table has fallen dangerously in many areas of the State of Iowa. It should be given careful study and an attempt made to so manage our water resources that they will either be restored, to some extent, or at least not be permitted to drop further, dangerously." Darling also insisted that the costs of such a study should be borne by agricultural as well as conservation interests. [11]

The former chief of the Biological Survey wrote to Governor Robert Blue urging that the streams in Iowa be preconditioned before they were stocked so that vegetation could be preserved and the new fish could be expected to survive. Darling badgered the governor to take a greater interest in appointments to the Conservation Commission and in conservation of the state's resources. Blue, in response, asked Darling to suggest names of persons who might serve on a fact-finding committee to gather data for the use of the Conservation Commission in suggesting changes to be made in Iowa's laws.[12] Darling angrily replied the same day:

I cannot resist the observation, however, that such a study and your acquiescence in it does not in any measure heal the breach caused by persistent and unethical interference with the work program of the Conservation Commission and their technically trained conservation officers. The Conservationists of the State would prefer to serve with you rather than criticize. The choice between the two alternatives is up to you.[13]

Finally, Ding unloaded publicly. He sketched a *Register* cartoon depicting Blue as a burglar removing the contents of the people's Conservation Commission "safe." Although Darling insisted that he did not mean to ignite a political issue, the cartoon started a firestorm of reaction across the state. The mail from assenting sportsmen flooded Darling's office. "The boys are thoroughly aroused," Darling wrote an Iowa publisher, "but they write to me instead of to the governor. If he got all the letters that I have received he'd think his coattails were on fire."[14]

Darling stood foursquare behind a proposal to establish a state Natural Resources Council that would assess the impact of developments within and adjacent to the state and would especially scrutinize the construction of various dams. Governor Blue had requested an opinion from his attorney general concerning the establishment of such a council. When the attorney general made his report, Darling attacked the governor again. He wrote that the report "denying you and the interim committee the right to set up a state Natural Resources Council," pending legislative action, was one-sided, fatuous and "contrary to your verbal commitment." Darling was also angered by arbitrary personnel changes in the field of state conservation and said so.[15]

Ding, who had pushed for years for a program that would teach conservation to teachers, also lent his hand to the establishment of the Iowa State Teachers Conservation Camp. At the end of the war CCC camps were standing unused in several of the state parks. At Springbrook, west of Guthrie Center, there were 1,400 acres of parkland within which the CCC had done some road construction and other work. The CCC had left behind a cluster of cabins and a main lodge. The site was ideal for continuing education in conservation; it had a stream, a marsh, a lake, some forest, and a prairie. Mrs. Addison Parker, a member of the Conservation Commission, requested and received Darling's assistance in making the project go. A determined conservationist, later a member of the Natural Resources Council and one of the first women ever to serve on the Conservation Commission, Mrs. Parker often went to Darling's office for advice. "I never went there but what Jay said, 'We need to teach conservation,' " she recalled.

The Conservation Commission staff taught at the Iowa State Teachers Conservation Camp the first several years of its operation. Later, faculty at Iowa State Teachers College (now the University of Northern Iowa) took charge of the program of instruction. Attendance increased each year and the facilities were improved. Fees were minimal, and the camp provided teachers with credit toward college degrees. After several successful years the federal government noticed the progress and approved a $400,000 grant that made permanent structures and increased stability possible.

Darling suggested a related use for CCC camps elsewhere in the nation. He wrote Ira Gabrielson, who had recently moved from the Fish and Wildlife Service to head up the American Wildlife Institute, to suggest a "camp for boys where they'd do some patrolling on wildlife refuges and get an education at the same time, as with forestry summer camps," especially in the Dakotas and Montana.[16]

The former Survey chief had seen changes in the making. He had earlier been uneasy about reports that Harold Ickes was about to resign. He wrote the secretary of the interior, "I have noticed at frequent intervals indications that you might be leaving the Department. That would be a tragedy. I hope you will stay as long as the Department exists and appoint your own successor." When Ickes did resign, Gabe announced to Ding that he would not have the job of secretary of the interior "for five times the salary they offer." Gabrielson said, "The Secretary went out of the Department with the loudest political explosion that has been heard around here in many, many years." When Gabe also departed, Ding wrote, "I was sorry that Gabe had to leave his job as Director of the Fish and Wildlife Service, but sorrier for the reason which made him do so. Gabe is not in good health. He hasn't been feeling himself for a number of years and I think it is due to the strain of that job." Darling's protégé Clarence Cottam was appointed assistant director of the Fish and Wildlife Service, and Al Day was named its director.[17]

One bright spot on the conservation scene was the success of the Cooperative Wildlife Research Unit Darling had christened at Iowa State College about fifteen years earlier. However, that too had its shortcomings for Iowa: "As a matter of fact up at Ames we have a special department to train wildlife technicians. Those men are snapped up by Texas, Pennsylvania, New York, Missouri, Michigan, Wisconsin and Minnesota and I don't know how many other states.''[18]

Ding had long been suspicious of the Army Corps of Engineers, and his cartoons reflected disdain for what he saw as its tendency to bottle up every river and stream in the nation. Even so, some individuals were unaware enough of his views to invite him to a meeting honoring the Corps. "Asking me to preside at a meeting to recognize the Army Corps of Engineers," snapped Darling, "would be like asking Senator Bilbo to perform a wedding ceremony for Rochester and Shirley Temple."[19]

Darling also took a jab at the Army for its postwar paranoia. He wrote Iowa Senator Bourke B. Hickenlooper concerning the Army's instinctive secrecy. Ding had visited with an atom scientist who had found it necessary to explain to army personnel the plain, everyday theoretical principles of atomic fission. According to Darling, "The Army's eyes popped out in amazement. They figuratively speaking, looked under the beds, up the chimney flue, and locked all the doors and pulled down the blinds at these amazing disclosures." "My God," said the scientist, "you might as well ask us to keep the Mississippi River a secret," according to Darling's account. "That stuff," Ding noted, "is in all the modern physics textbooks in the world."[20]

Darling met Winston Churchill shortly after the former Prime Minister made his famous "iron curtain" speech at Fulton, Missouri, and Darling's friend and the perennial adviser to American presidents, Bernard Baruch, subsequently coined the term "cold war" in a South Carolina speech. But the war uppermost in Darling's mind was the one to win public support for conservation of the essentials of human life. He wrote of his concerns: "Two World Wars, in which we have fed and munitioned the world, have made a tremendous hole in our Natural Resources. We have dug into our iron ore until it is about gone. Our forests are depleted. Our oils are now a scarcity and so far as soils and water resources are concerned they are worse off than any of them—only we don't know it."[21]

Darling was pessimistic, and he was critical of policies aimed at feeding the world's hungry from America's pantry. To his friend Baruch, Ding wrote, "Geoffrey Parsons and Helen Reid, of the New York *Herald Tribune*, are converts to the theory that there is enough for all and more. Their editorial policies support that out-moded theory. They even cut out my cartoons which bear on that subject. The same thing is true of most of the editors and commentators who reach the public eye and ear."[22] The conservationist worried that a biblical adage was being interpreted literally by a population naive about the finite supplies of the fuels of life:

The trouble with me is that my convictions run counter to some of the tendencies which are now being broadcast that we should attempt to feed the world. You might as well try to set up an organization to feed all the Kansas grasshoppers and the result would be just about the same—you would just succeed in getting more grasshoppers. . . . There comes to my mind the old adage that casting your bread upon the water will return it fourfold. The only trouble is that in this feeding of the starving peoples of the world, we get fourfold return of population pressure rather than fourfold return of our bread.[23]

The trend of American social legislation and international policy had swerved decidedly away from Darling's position. His views were considered reactionary by some of the editors who subscribed to the *Herald Tribune* syndicate and by members of the staff of the *Register* and *Tribune* as well. His support of the principles by which he had lived and been rewarded was apparently losing its currency. In 1946 he wrote, "You say there is no newspaper in Iowa which is making a fight for the American way of life. Why limit it to Iowa? . . . That is the tragedy to me that nowhere except a few industrialists talking to themselves is the case for the American way of life and freedom of individual initiative heard in the discussions." Darling, who was nearly seventy, claimed newspapering in general "isn't nearly as much fun as it used to be." He reminisced, "Getting out the old Sioux City *Journal* was a daily picnic and in the early days of the *Register* it was much the same."[24]

Near the end of 1947 Darling left his drawing board to do an ecological study of the islands of Hawaii as a consultant. As he departed, he was fairly optimistic about at least two conservation matters in his state. For one, his friend Ira Gabrielson had been secured to conduct a study of conservation management in Iowa, and Darling had confidence that Gabe would bring in a beneficial and truthful report. For another, Ding had confidence in the chairman of the Conservation Commission and in his determination to operate it free of political interference.[25]

The Hawaiian tour of duty became a labor of love for Darling, and he spent longer than he had planned in turning out a report more detailed than had been anticipated. Even so, when he returned to the United States, he was in plenty of time for the 1948 elections. He shared the general Republican optimism on the political front and, fresh from his labors in Hawaii, he took a more optimistic view of his own activities: "I never had the time to spare for conservation either but as I look back over the years, my labors in that field give me more real satisfaction, even with the little accomplished, than anything I have done in my life."[26]

He threw himself into an assignment for which he was well equipped. Ries Tuttle, outdoor editor for the *Register* and former president of the Iowa division of the Izaak Walton League of America, requested Ding's help in drafting resolutions for the state League meeting. Darling did just that, and in the proposed resolutions Governor Blue was criticized for low salaries in the Conservation Commission and for retarding programs the legislature had approved and funded.[27]

Darling's health had not improved. He wrote Harvey Ingham, former editor of the *Register,* apologizing for missing his ninetieth birthday party: "I hope you don't mind my failure to show up at your birthday party and assure you that my presence or absence could have had no possible relationship to my high esteem and personal affection for you. I never go to parties. . . . I'm getting so old and deaf that I can't hear anything and I never know what to say or do when I get there. . . ."[28]

"Dr. Darling" continued to practice medicine on himself, however. He commiserated with an acquaintance who also suffered from asthma: "I keep mine pretty well under control when I don't smoke too many cigarettes or drink too much gin. . . . I find that the proteins are likely to poison me and bring on spasms of bronchial asthma. Sea food is the worst but of all things plain, ordinary farm butter (which is least suspected by the medics) was one of the offenders."[29]

As election day neared, Darling lamented the political impotence of his old friend Henry A. Wallace, who had been dumped by the Democratic candidate Roosevelt four years earlier and was running for president on the Progressive ticket against Harry S. Truman and Thomas E. Dewey. Wallace had lost credibility in the atmosphere of the cold war with his one-world philosophies and reputation for mysticism. Ding commented, "No one around here is for Henry Wallace. Even his old associates dodge the mention of his name and his campaign. In fact, he is so weak that I'm afraid his influence upon the vote for Truman will not be as effective as we originally thought it might be."[30]

Darling avoided the political conventions. He had covered every national party convention from 1908 through 1944 but "whatever thrill there may have been in the older days has completely worn out for me," he wrote. Darling, with many others, had anticipated the final result of the election and had drawn a cartoon for the syndicate showing Truman and Wallace being booted down the steps together. The *Herald Tribune,* under the hand of Helen Rogers Reid, had supported Dewey. She wrote Darling following the election: "We expected to use your enchanting cartoon . . . before Tuesday evening was half over, and Geoffrey [Parsons] had written a wonderful victory editorial. Work in the *Herald Tribune* office ended at dawn Wednesday morning and I felt a good deal as if I had taken a physical beating.''[31]

The election results had delivered a body blow to Darling too. He was tired and hurt. Shortly before the election, Ding had written that he should have retired a long time ago. He hesitated, though: ''I guess retiring is something like getting ready to die—you are never quite willing to face it, at least I'm not and I can't figure how life would be very interesting without a mountain of work which I've got to climb over.'' By the spring of 1949, however, Ding had decided to cut bait. His retirement was announced in the Des Moines *Sunday Register* April 24, 1949. Tom Carlisle, who had worked with Ding since 1926, began the regular production of *Register* cartoons. *Newsweek* magazine devoted nearly a page to Ding and his successor, described by the *Newsweek* writer as a slight, quiet zoology student who had tired of stuffing a moose when Ding introduced him to cartooning.[32]

There had been interruptions in Darling's nearly fifty years as a cartoonist. He had tried to retire in 1943, although he returned to work following a brief absence. His health had continually forced him onto a reduced schedule. He had been unable to work during his repeated and extended hospital stays. Even so, the prolific Ding had probably drawn 15,000 or more cartoons for the *Register;* the *Herald Tribune* syndicate; *Collier's* magazine; and occasionally for *Better Homes and Gardens, Successful Farming,* and other publications. He was calling a halt to the vocation that began when as a youngster he tried to copy a postcard addressed to his father.

Darling was not planning to step into a vacuum, though. His fight had not been won. Perhaps it never would be. He was stepping into combat with waste and indifference. It was going to require much of his aging and damaged body and his agile mind.

# 18

## *Homogenized Conservation*

SHORTLY AFTER Ding retired as a cartoonist, the Izaak Walton League passed a resolution urging the Department of the Interior to place Darling's portrait on the 1950–51 migratory waterfowl hunting stamp. Darling was aghast. He wrote Secretary of the Interior J. A. Krug, "Please oh please, Mr. Secretary, DON'T." Darling's foresight told him nothing would bring more irritation and ridicule upon future secretaries of the interior and officers of the Fish and Wildlife Service than to start a campaign of recommending portraits instead of migratory waterfowl on the duck stamp. He concluded his letter, "Since it is I who have been chosen for this innovation I want to be the first one to protest against it."[1] Ding prevailed.

Darling maintained his fervid interest in conservation matters at the national level and in issues more substantial than questions of duck stamp design. One of the most informative events in his young retirement was a trip to Alaska, where he viewed some of the results of his reign as chief of the Biological Survey. As chief, Darling had been responsible for transplanting a small family of musk-oxen from Lapland to Nunivak Island, where the musk-ox had become extinct. Darling found that the animals had become completely antisocial and that visitors to the island refuge were likely to have to scurry for their safety. He whimsically suggested that "we put some of our recalcitrant grizzly bears on that same island and see if a little real competition won't sweeten their dispositions."[2]

His antidam instincts were reinforced when he observed the changes on the Columbia River. In spite of refutations from the Department of the Interior and the dam builders he wrote, "The twelve-million-dollar run of Salmon up the Columbia River is thinning like the ranks of the Grand Army of the Republic." Darling further suggested that the hydroelectric plants powered by the water behind the dams "will have to work overtime to show an annual net profit of twelve million dollars a year."[3]

Darling felt mixed emotions on timber cutting in the northern forests. On one hand, he suggested that the operations of the lumbermen were not as bad as had been reported. Selective cutting of timber had been given up, he noted, for alternate square-mile cutting. Young trees had been crushed by the felling of larger ones, and the dead trees and underbrush presented serious fire hazards. Even though he favored square-mile cutting, Darling was suspicious of lumber company proposals that they be allowed to cut timber in the national forests and the national parks. "I hope our forest custodians don't fall for that baloney," he wrote.[4]

As part of his Alaskan trip Darling also spent time in the northwestern United States, where he found reasons for both satisfaction and distress. Several members of the former Biological Survey staff were still in the field, and Darling was quite delighted to find that the one-time resentment of the old-timers in the Bureau over his violent housecleaning methods had been considerably replaced by what they now thought of as pleasant memories of accomplishment. Darling was worried, though, about the burgeoning of the elk herds in the state of Washington and, in particular, on the Olympic Peninsula. There were, he reported, five or six areas of elk overpopulation in Washington that no one seemed to have the courage to handle. Darling believed that the herds would have to be reduced; and he reported to Archibald B. Roosevelt, son of Theodore Roosevelt, that the experts estimated the huge herds could be cut by more than eighty percent without any danger of their becoming extinct. Roosevelt, a leader in the Boone and Crockett Club created by his father, had a special interest in the matter. Near the turn of the century, the club had worked to save the elk from extinction in the Jackson Hole area of Wyoming. In the days when members of the Benevolent and Protective Order of Elks wore elk teeth on their watch chains, tooth hunters had nearly wiped out the herds.[5]

Although Darling returned from Alaska and Washington with some new cares and concerns, he enthusiastically endorsed the attitudes and energies of the U.S. Fish and Wildlife agents all along the line of his journey. Reminiscent of Darling's days as chief of the Biological Survey, "they are all working like beavers and [are] only impatient at the limitations which are put in their way," he reported.[6]

When Darling ended his cartooning career, he turned his considerable talents and resources to conservation problems and, in particular, to methods of conservation education. He had become convinced that unless the study of land, water, and vegetation could be ecologically stirred into school curricula, the future of conservation, and thus the nation, was going to be bleak: "Turn the natural resources of any area over to an ecologically ignorant populace and ecologically ignorant leaders and they will rape the land and waters with as little regard for future consequences as the profit-motive boys display. . . ."[7]

The concept of an interdisciplinary conservation education was practical, he insisted. "If whole milk can be homogenized so that the cream will not separate from the milk, then certainly nature's laws governing natural resources, which are an integral part of every branch of the knowledge by which we live, can be fused together so that they are inseparable."[8] Darling contended that conservationists, with their heavy publication programs, were "manufacturing rifle ammunition, which does not fit the educational shotguns." Although many good conservation publications were being produced, they were circulated on the false assumption that teachers were willing or able to add conservation courses to existing curricula, he claimed. Any improvement would have to come in the rewriting of textbooks in history, political science, biology, botany, chemistry, geography, sociology, and other subjects. "All are taught in a manner equally oblivious of the fact that conservation of natural resources is a major factor in our national economy, standards of living, industry, international relations, and, finally, peace and wars," Darling fumed.[9]

Even though his own college career was an uneven one, Darling espoused higher education, research, and the scientific method. He formed many of his opinions according to scientific findings and occasionally put his own reputation in the hands of scientists. The most worrisome aspect of higher education to the direct and active Ding was the danger of doing research and finding answers that would never reach the practitioners who needed them. He was also critical of the specialization required of advanced degrees in the social sciences to the exclusion of the life sciences. "One of the major fallacies in all our public utterances and educational systems," he once wrote, "is that sociological and economic doctors never studied biology." When the makeup of the state Natural Resources Council was under discussion, Darling argued with a proposal made by the Iowa State Education Association calling for a preponderance of lay members. Darling urged that scientifically trained members were to be preferred. There was "less chance of Iowa being misdirected," he said, "if the final interpretation . . . of the deciding body . . . is made up of highly qualified scientific minds rather than lay opinions."[10]

The cardinal precept of Darling's educational philosophy was that teachers had to be taught conservation if the students were ever to be exposed to the interdisciplinary approach he believed necessary. Ding's spirits therefore rose when he became part of a project sponsored by the American Association of School Administrators. He was named a member of a committee whose aim was to compile a series of articles and texts for a 1950 yearbook devoted to conservation. Darling explained that the organization was affiliated with the National Education Association and suggested "it is rather a powerful group or I wouldn't be giving my time to it." Approximately eight months later, however, the former cartoonist resigned his membership on the conservation yearbook commission. The yearbook proposal had drifted toward requiring the addition of conservation courses to school curricula. Darling became frustrated. He suggested there was easily a five-foot bookshelf full of better books on conservation than the draft prepared by the commission. "I would be embarrassed to have my name connected with the project," he wrote.[11]

The practical-minded Darling detested the publication of scientific research for its own sake, and he saw as pointless the accumulation of education credits without practical applicaton of what was gleaned from the curricula. He earlier wrote Aldo Leopold, "I get so damned tired of education being used as an alibi for more energetic methods that there is a strong temptation to ridicule the whole educational system and its methods. . . . I'm all for education, as of course you are, but it is by no means a cure-all for wasting resources, or international relations or rape."[12]

For that reason and others, Darling was proud of what he had initiated at Iowa State College with the 1932 creation of the Cooperative Wildlife Research Unit. Although he occasionally grew impatient with the unit's activities, he repeatedly held it up as a shining example of serious research and teaching put to practical use.[13] When Carl Drake, a professor at Iowa State and a member of the wildlife unit faculty, was not named a member of the Natural Resources Council, Darling was disappointed but not discouraged:

Iowa State College—has the best nucleus of scientists concerned in the wildlife field of any educational group in the state . . . no matter what biologist might be appointed for that place in the State Council the men at Ames were so thoroughly interested in all branches of wildlife and fishery conservation that they would see that any biologist would hear from them on methods and personnel essential to work on the research and reports.[14]

Darling dealt with hard facts untempered by sentimentality or religious philosophy. A warm and responsive human being with a gift for bringing out the best in more reserved persons, Ding nevertheless looked upon the world about him through a pragmatist's eyes. Even though he was the son of a deeply religious father and well-versed in the Scriptures, Darling did not interpret the Bible's words literally. In assessing a book on population and resources, Darling wrote:

Ever since that story got around several thousand years ago that Samson killed a thousand Philistines with the jawbone of an ass, hard naked facts have had a difficult time trying to catch up with false rumors. . . .

Anyone who has recently tried to feed a family of four plus a couple of unexpected guests on fifty cents worth of lamb chops will keep his fingers crossed while swallowing that story about feeding the multitude with only five loaves and two fishes and picking up twelve baskets full of leftovers.[15]

Darling's appreciation for the scientific approach was apparent in his conservation-related activities and extracurricular interest in and experimentation with plants and wildlife. Nowhere, however, was his personal application of scientific observation and the empirical system more apparent than in the case of the "red tide." His first experience with it was in December 1946. Darling was at his Captiva hideaway, when fish nearby in the Gulf of Mexico began dying mysteriously in large numbers. The stench of decay as the bloated dead fish were washed ashore made Captiva and Sanibel virtually uninhabitable. It was a threat as well to the local, tourist-oriented economy. Darling and others called upon the Division of Fisheries in the Department of the Interior for scientific assistance to find the cause of the fish deaths and to propose a remedy. In Darling's opinion, the response from the federal government was inexcusably tardy and insufficient. He wrote to Oscar Chapman, under secretary of the Department of the Interior:

If any individual, corporation or alien enemy had threatened the health, destroyed the economy, food resources and employment of as many families as has resulted from the so-called "Red Tide" on the west coast of Florida, every Department of Government including Congress and the Army and Navy would have been up in arms. Hundreds of millions of food fish have died, practically all marine life is extinct over a 75-mile stretch of these coastal waters, and the total consequences in economic losses can hardly be estimated at this time. . . . The Division of Fisheries, official custodian of marine life, has participated in this emergency to the total extent of two letters and a quite erroneous diagnosis released from Washington to the Press.

At his own expense, Darling had made arrangements with Miami University to send scientists to the scene to observe and investigate. As a result of their findings and his own the conservationist concluded that the Interior Department had been in serious error; the cause of the fish mortality was not the red tide of the East Indies but a yellow tide, a strange and new microorganism.[16]

Darling backed up his claim with personal observations, which he and Carl S. Miner, another island resident and head of the Miner Laboratories in Chicago, committed to paper. Darling noted that December 27, 1946, a patch of yellowing water about 300 feet in width and 400 to 500 feet in length was spotted less than a mile south of the Sanibel Light about twenty miles west of Fort Myers. He described conditions in the midst of the yellow water: "Fish were noticed arising to the surface, gulping for air, turning over on their side and sinking. That caused the observers to look into their own baitwell in the boat where they had confined a number of pinfish, several jackfish (which are particularly hardy) and two or three small mullet. The fish in the baitwell had all come to the surface and were gulping for air. Later they all died."

Ding returned to Miner's residence with samples of the yellow water. When the liquid was run through filter paper, it left a bright yellow ring. Samples of the yellow material were put under a microscope and "it was immediately discernible that a live and very active organism was predominant and gave color to the water." The retired cartoonist put his skill to work by drawing several detailed sketches, from various perspectives, of the organism beneath the lens. In scientific terms Darling described its movements, the structure of its body, and the locations of its flagella. When the organisms died, they turned a rusty color. No identifying information could be found in available reference books.

Miner and Darling observed that, with the presence of the yellow tide, residents of the islands suffered persistent coughs. When the wind died, however, the coughing epidemic "miraculously" stopped. Darling experimented further. He placed a pan of the yellow water over a flame and it emitted such a quantity of irritating gas that observers were driven from the room. Finally, Darling and Miner devised what they believed was a remedy: "A crumb of copper sulphate the size of a small pea was pulverized and dusted over a twelve-quart bucket of water heavily infested with the organism. Within a few minutes all the organisms were dead and sank to the bottom of the bucket, depositing a reddish-brown sludge."[17]

Tragically, in 1952 the mysterious plague struck again. Darling grieved:

Dead fish by the hundreds of thousands are washing ashore and the stench is beyond belief. Five years ago we went through this same sort of blight and spent most of the winter burying dead fish that washed up on the shore, and we don't feel inclined to stick it out through another season of such extreme discomfort. . . . All the inn-keepers and hotel men along the west coast of Florida are hysterical in their desperation over the situation and a good many of the early arrivals are already packing up to leave.[18]

The Darlings, too, left the islands that winter in favor of Spain.

Ding lamented the lack of authority given subordinate installations by their parent government agencies. In a letter to Milton Eisenhower, with whom Darling had worked in Washington, D.C., he wrote of the red tide incident: "A man with a tin cup and a microscope sitting any where on the hundred-mile stretch of beaches, from Boca-grande south, could assemble more scientific information than did the several hundred-thousand dollar equipment and staff of the Galveston Research Laboratory." He also wrote his good friend Clark Salyer, employing a typical Ding medical metaphor: "Five years ago I watched the so-called Red Tide generate, come to flower and burn itself out in about four months' time—to be exact the latter part of February 1947. The Fish and Wildlife Service got a man on the job sometime in March. He was a good doctor but the patient was already dead. . . ."[19]

There were other problems in paradise, and Darling did his best to keep abreast of them. When he heard from an old and trusted friend that commercial crawfishermen had been allowed to invade the waters of the wildlife refuge at Dry Tortugas, stripping it clean of crawfish, he wrote in anger to the Fish and Wildlife Service. There were heated exchanges. "I wonder," Darling wrote, "if the Fisheries Division of the U.S. Fish and Wildlife Service is still a dead oyster."[20] Darling was also dissatisfied with a response he received from the Park Service:

I really had to laugh at it, it was so very sanctimonious and unctuous in its protestations. . . . The Report reminded me of similar documents I used to get from the Coast Guard and local administrators along the eastern coast of Maryland. . . . You would have thought from their combined reports that no one ever shot a gun out of season in that neck of the woods and that everybody went to Sunday School seven days a week.[21]

When a refuge was finally set aside on Sanibel and Captiva, Darling influenced the appointment of W. D. (Tommy) Wood as its manager. Darling had great respect for Wood, who was also responsible for supervision of refuges along the western coast of Florida from Cedar Key to and including Key West and used an Interior Department amphibious airplane to make his rounds. Through Wood, Darling kept a finger on the pulse of the islands even when he was in Des Moines. By fall of 1952 island "improvements" were troubling Wood. The land was being cleared, a yacht basin had been dug and walled, and the spoil had been used to fill swampland. Wrote Wood, "I personally hate to see it come to this."[22]

Years earlier, Darling had been introduced to the tiny Key deer under tragic circumstances. He was aboard a launch cruising among the mangrove thickets of the Florida Keys, when he saw smoke rising from one of the islands. He was told that fires were set on the island to flush the toy breed of deer onto the beach where they could be shot for meat. Darling was moved to draw an evocative sketch then and there on a lapboard as he sat in the boat. The hunters had nearly wiped out the unique deer herd. Darling was eager to see the slaughter halted. Florida conservation officials were preoccupied. Nothing was done at the time, but the Key deer sketch played an important part in whipping up public support for a bill calling for a federal Key deer refuge. The cartoon was widely circulated by proponents, and in 1952 "Save the Key Deer" became the theme of National Wildlife Week, sponsored by the National Wildlife Federation. With such public support and years of effort by James Silver, southeastern regional director for the Fish and Wildlife Service, and Pink Gutermuth's North American Wildlife Foundation, a Key Deer Refuge Bill was passed and signed into law in 1957. Further assistance, including financial aid provided by the Boone and Crockett Club, made it possible to purchase lands needed for the refuge.[23]

Darling's scientific and pragmatic approach extended to the nation's inland waters as well as Florida's coast and to the dams he believed were a blight on the natural resources of water, land, and vegetation. Darling distrusted the Army Engineers, and despite his great respect for Herbert Hoover, Ding had one reservation about him— Hoover, too, was an engineer: "In spite of my one-time close relationship to Herbert Hoover, I never was able to penetrate the boney shell which seems to encase the minds of all engineers. . . ." The Hoover Commission, established by Congress in 1947 and headed by the former president, made its recommendations for government reorganization in 1949. Hoover's commission had benefited, Darling asserted, from the views of "some men who at least could detect the waste and destruction which was bound to follow in the wake of this present extravagant program" and it recommended that the river, harbor and flood control work of the Army Engineers be transferred to the Department of the Interior.[24]

Ding was not against dams altogether, but he contended that each proposed dam should be carefully studied and its environmental effects should be weighed. To his colleague in the dam-fighting business, Elmer T. Peterson, Darling wrote, "I liked your piece . . . about the difference between dams constructed in the mountains to catch flood waters and those built on the prairies which will be short-lived from siltation and of no particular value for any purpose except to spend money to re-elect some shyster congressman." Siltation distressed Darling for several reasons. It soon filled in the bottoms of the reservoirs the dams were supposed to provide. It made water dirty and unsuitable for the recreation that was often another rationale for dam construction. It smothered underwater vegetation, making it impossible for any but the roughest fish to survive. Finally and most important, however, Ding suggested that "the top soil which goes swirling by in our rivers at flood stage may look like mud to you but it is beefsteak and potatoes, ham and eggs and homemade bread with jam on it. . . ."[25]

Darling made another forceful objection to the big dams on the nation's rivers:

And speaking of the losses of the homes and property of people who build too close to the shore of the Great Lakes, the damage to them is microscopic compared with that of the thousands of acres of farm land, farm buildings, farm families and actual food production being destroyed by the submergence under the reservoirs back of the big dams already built, now in the process of building, and on the schedule of the Army Engineers for future construction.[26]

He had what he thought was a better idea, backed by experience, knowledge, and successful demonstration projects. In a hearing conducted in 1950 in conjunction with plans to build Iowa's Red Rock Reservoir, Darling testified, "We have ample proof on demonstration areas that runoff can be stopped before the waters reach the rivers and thereby save not only the water but the soil which is washed off with it. On such demonstration areas we have the triple benefit of flood control, soil conservation and restocking of our subterranean water table."[27]

Although Darling was occasionally cheered by an isolated development in the conservation of resources, he was generally depressed at the outlook. "We have less of everything than we had fifty years ago," he observed. The affairs of Iowa conservation, he feared, were falling back into the hands of the politicians. He decried the clash of personalities in the U.S. Fish and Wildlife Service where, he had heard, Al Day had "turned" on Clarence Cottam. Commenting on the success of the bird refuges created in the 1930s, Darling wrote, "I wish somebody would start a few refuges for us humans who are beset by the political predators and threatened with extinction if we're not there already."[28]

Ding even suggested that he was losing his knack for making a gripe sound funny. Equally uncharacteristic, Darling was favorably impressed with the work of the National Council of the Congress of Industrial Organizations in analyzing conservation from the point of view of organized labor. Darling the Republican accepted hopeful signs for conservation wherever they could be found. In 1952 he noted that Adlai Stevenson, the Democrats' presidential candidate, "is the only man who has expressed any intelligent interest in the cause of conservation." He also questioned the depth of President Dwight D. Eisenhower's devotion to conservation.[29]

Darling remained convinced that the one best means of combating the slippage in conservation was widespread education through a conservation information bureau and the popular press. He was pessimistic about the prospects for his news bureau idea because, "no one seemed interested in a Clearing House unless they could be 'it.' " Darling suggested a direct route to the press. He proposed that the Friends of the Land issue an invitation to a group of newspaper and magazine leaders in public thought to attend a dinner at the Waldorf-Astoria Hotel in New York where a deluxe show could be put on by a series of spokesmen in specified areas of conservation. Darling suggested the idea to Bernard Baruch and "he was all for it—even offered to give the dinner." Still, the dinner had not occurred, and Darling was disappointed: "I wish somebody would undertake that sometime because I honestly don't believe that the men who control the editorials and policies of the great publishing houses really know anything about the fundamentals of our conservation principles."[30]

Darling's discontent with his own efforts did nothing to reduce the hero worship he received from others interested in conservation. The Des Moines women's Izaak Walton League became the Ding Darling chapter. The Conservation Commission named a small lake on Honey Creek, north of Fairfield, Iowa, for Darling as well. Darling noted that he would always "blush a bit" when writing the Ding Darling chapter. He also warned the Conservation Commission of the danger in naming lakes for conservationists: "You see some day you may have an outstanding conservationist in this State by the name of Pete Specknoodle or even Adolph von Hausenblausen and what the hell would you do then when you have to follow the precedent which you have set in naming a lake after me?" Vanity, Darling claimed, could find justification for nearly anything, however, and "I begin to see some virtue in the name 'Darling Lake on Honey creek'. . . ."[31]

Darling accepted such acclaim and recognition with reserve because he saw far more to be done than had been accomplished. He had built a reputation as the "best friend the ducks ever had." In his ecological view that tagline represented a failure to make the connection between healthy human beings and other forms of life. Near the end of 1953 he put his concern in the vernacular for his friend Bill Hard, a roving editor for *Reader's Digest:*

Of course you understand that I am not nearly so much interested in the preservation of migratory waterfowl as I am in the management of water resources and the crucial effect of such management upon human sustenance. Wild ducks and geese and teeter-assed shore birds are only the delicate indicators of the prognosis for human existence just as sure as God made little green apples.[32]

Darling was nearly seventy-seven years old when he wrote those words, but his mind would not rest. He saw much to be done, and he planned to keep slugging away.

# 19

## *Losing Ground*

BY EARLY 1954 Darling was devoting more of his time and correspondence to conservation issues. That year he was in the thick of battling efforts of the Eisenhower administration to construct Echo Park Dam in Dinosaur National Monument, which was on the Upper Colorado River in Utah and one of the prized units in the National Park System. Darling wrote Douglas McKay, Eisenhower's secretary of the interior, protesting the department's violent reversal of a sound national policy. Darling railed against what he called the unnecessary flooding of a prime natural resource and put its prevention at the top of his list of priorities. Even though he had been concerned with Communist infiltration of the federal government, dating back to his days in the New Deal administration, Ding objected to Senator Joseph McCarthy's 1954 televised government loyalty hearings because McCarthy and the subcommittee were "hogging the limelight" and distracting editors who might otherwise have paid serious attention to the Dinosaur National Monument crisis.[1] Darling wrote:

Even the Huns learned not to bomb the great accumulation of art objects in the big museums, art galleries and cathedrals. They even hid the masterpieces of the painters in the salt-mines so that they might not be destroyed but our nation continues to ruthlessly whack away at the spectacular monuments which nature has designed[,] and this Echo Park Dam is so very unnecessary and completely reverses an established national tradition of many years.[2]

That August, Darling participated in an eightieth birthday celebration for Herbert Hoover at Hoover's birthplace in West Branch, Iowa. Darling was not enthusiastic: "It's pretty hot here for an all-day celebration and they've built up such a busy schedule for Hoover—including a major speech and one from Vice President Nixon, lunches, a

visit to the old homestead, the ol' swimming hole . . . that I'm afraid
they'll kill the old boy off.'' Darling joined in, however, and made the
most of his opportunity to chat with the former president about the
Dinosaur Park problem. Hoover assured him that the Dinosaur Park invasion was not a dead issue. He urged Ding and his comrades to "keep
shooting" because, unless there was a shower of individual protests to
the various members of Congress the Echo Park Dam would probably
be authorized. Darling responded with a volley of letters to the Iowa
congressional delegation. He condemned the project and suggested that
every known authority on the subject of conservation had also condemned it, "with the one exception of the promoters and the acquiescence of the U.S. Army Engineer Corps."[3]

In the midst of doing battle for conservation, Darling expressed
dismay at the performance of President Eisenhower. The former cartoonist was not pleased with Ike's choice of Douglas McKay for secretary
of the interior, and he was mildly miffed at the lack of access accorded
conservation interests by the president's brother Milton, on whom, Darling wrote, "I suspect Ike leans for conservation advice."[4]

In the Dinosaur Park question, Darling saw a dangerous capitulation on the part of the Park Service and a failure to fulfill its obligations
as spelled out in the 1916 legislation that established it. He saw
capitulation by the Park Service and its subjugation to the will of the
concessionnaires as well: "Production-line bus rides, cheap souvenir
stands and juke box entertainment are now offered as a substitute for
the quiet beauty of our old nature trails, the wilderness environment
and wildlife habitat for which our national parks were originally set
aside."[5]

Darling, who played a major role in the creation of the duck stamp
program, was outraged at a movement in Congress aimed at repealing
the Duck Stamp Law. "We tried to get Congress to appropriate the
necessary funds for about twenty years," he stormed, "and that is why
we inaugurated the Duck Stamp provision in 1934." In twenty years
duck stamp sales had brought in approximately $29 million on the
pretext that the money would be used to aid the propagation and conservation of ducks and geese, Darling recounted. Congress, however,
had impounded millions of those dollars and refused to release the
funds for migratory waterfowl management.[6] Darling wrote Iowa
Senator Bourke Hickenlooper requesting the lawmaker to use his influence on Interior Secretary McKay, Agriculture Secretary Ezra Taft
Benson, Commerce Secretary Sinclair Weeks, and other members of the
Migratory Bird Commission, urging them to release funds for the purchase of additional refuge areas. The proposal was not for new refuge

areas, Darling emphasized, but for the completion of the original plan to buy sanctuaries piecemeal, as was economically feasible and necessary. Darling argued:

The prices are too high—sure! So are all prices. But the particular lands now in question, and their objectives, remind me of the little airfields built twenty years ago when we had not yet learned of B-29's. . . . These . . . projects will cost in the neighborhood of $300,000, which is chicken-feed compared to the millions the sportsmen have kicked in toward wild waterfowl management. I don't have the exact figures at hand but since the Duck Stamp law was passed in 1935 the U.S. Government need not feel too extravagant in rounding out the refuges already established.[7]

Although the prices of land were high, prices of agricultural products were low; and Secretary of Agriculture Benson was swamped with agricultural surpluses. At the same time, the Department of the Interior was rationalizing the construction of dams by suggesting the water stored behind them could be used to bring arid land into agricultural production. Darling highlighted that irony and his attitude toward public power in a message to Secretary McKay: "As long as Secretary Benson remains smothered in surplus butter and other agricultural surpluses I see no virtue in adding more land for production by means of irrigation, and as to the production of electric power—that project is contrary to every instinct I have about American free enterprise."[8]

Darling's discontent with the Department of the Interior only deepened with the announcement that his friend and respected colleague Clarence Cottam was leaving the Fish and Wildlife Service. Darling wired Cottam, "I know of no man in the service . . . who has rendered more valuable services in the field of conservation or whose talents have been less rewarded."[9]

He later expressed his resentment to Cottam, who had been appointed dean of the College of Biological and Agricultural Sciences at Brigham Young University. Darling reiterated his disappointment in the Eisenhower administration and especially the Department of Interior:

I marvel that with all our years of emphasis on conservation and greatly increased national interest we can still pick our Secretaries of the Interior from men who have no basic convictions and practically no knowledge of the fundamentals of water management, wildlife and the other natural resources. In all the years within my memory there has been just one who achieved a fair degree of understanding and that was Ickes, who acquired most of his conservation education after he got into office. . . . I used to find in Milton [Eisenhower] the most practical and effective approach to Henry Wallace and

Rex Tugwell, when I needed to educate both the Secretary and Rex on the logic of some of the things we were trying to do at that time. Milton doesn't commit himself and usually has his Secretary answer my letters. Through Leslie Miller, ex-Governor of Wyoming and now on Hoover's subcommittee on Water and Land Management, I now find the only wire open to the Administration. Leslie Miller is okay but he finds it difficult to boil down his messages to one page— which is all that Eisenhower will read on any subject. God help us![10]

Ike, like Hoover, leaned toward confidence in the Army rather than scientists in the government bureaus, Darling complained. "To hell with the Army Engineers!" he exploded, "They are both ignorant and dishonest, but you'll have to admit that they are a damned sight better when it comes to propaganda in their own interests than are all the conservation organizations put together." Darling tangled with a combat branch of the military, too. The U.S. Army was poised to "grab title" to a large part of the Wichita Mountains National Wildlife Refuge in Oklahoma for use as an artillery range for troops at Fort Sill. The area constituted the last remaining habitat for the largest specimen herd of American buffalo, the giant bronze wild turkey, the Texas longhorn herd, and other vanishing species of the old western plains. Darling, who had taken a personal hand in making the area a wildlife refuge when he was chief of the Biological Survey, was outraged.[11] He was so infuriated that, despite earlier refusals to do so, he prepared a cartoon for publication.

In the six years following his retirement he had steadfastly withheld his talents. In response to a request made of him in 1951 he wrote, "I haven't drawn a picture for over two and a half years and I don't expect to make any more except in a case of proved necessity." He later observed, "My days of drawing cartoons and illustrations are over." By 1953, he was firm: "I wouldn't like to lower my standard by making any illustrations for publication ever again."[12]

The Fort Sill cartoon, titled "Speaking of Armed Aggression," was circulated to newspapers throughout the country. It pictured artillery projectiles raining down on the wildlife refuge, sending helpless animals and birds scurrying for their lives.

On the heels of the Fort Sill controversy, Darling learned the Navy wanted to add 655,000 acres to its Sahwave Mountain gunnery range and 1,372,000 acres to its Black Rock Desert area, both in northwest Nevada. Darling wrote Secretary of the Navy Charles S. Thomas:

Combining the Army and Navy's preemptory demands, without so much as "by your leave" from any of the other interested Federal authorities, you have picked out three of the most prolific and valuable areas which support traditional species of American wildlife left on this continent. Some of the species have been brought back from a vanishing fringe to a stabilized population at great cost in money and many years' effort. Primitive game ranges, unspoiled by industrial or agricultural invasion, are very rare and there is no place else for these spectacular species to which they may be removed and maintained in the wild state.[13]

Darling felt victimized by the demands of the military, and his suspicions were confirmed in a luncheon with Senator Hickenlooper, who informed the conservationist that "if anyone used 'national defense' as an excuse for their projects, no one—in the government or out of it—was well enough informed to raise any objection."[14]

The wheels seemed to be falling off the conservation wagon. In the area of game-law enforcement Darling was dismayed to learn that the Department of the Interior had issued permits to shoot protected birds. While in Florida he had talked with a cattle rancher on whose land the sandhill cranes wintered each year. Darling asked the farmer how the birds were getting along. "Hell," he said, "I have to shoot the damned things who keep eating up my corn. They just follow the planter right along the row and eat up the corn as fast as I plant it." "You can't do that," Ding said. "Sandhill cranes are protected." "Can't eh?" said he. "I got a permit from the Department of Interior."[15]

Darling, tongue firmly in cheek, wrote Interior Secretary McKay: "Why not make the corn surplus and the Sandhill scarcity cancel each other by you expanding the production of Sandhill Cranes and Ezra [Taft Benson]—God bless him—using the increased population of Sandhill Cranes to eliminate his corn surpluses?" Darling grew more dissatisfied with McKay's administration the longer the secretary remained in charge of the Department of the Interior. He wrote to Tommy Wood and his wife Louise: "I fear that our Secretary of the Interior, Mr. McKay, is guilty of criminal ignorance and he doesn't know that he doesn't know anything about wildlife management."[16]

In 1956 Eisenhower appointed a new man, Frederick Seaton, to succeed McKay, and Darling took hope from the change. To Clark Salyer, who remained with the Fish and Wildlife Service, Darling wrote, "I hope . . . that the new Secretary of the Interior proves a beneficial change." By September of 1956 Darling was more comfortable about voting Republican. Eisenhower was scheduled to drive through Des Moines, and Darling watched with some of the same detachment with which he had witnessed the celebration of the end of World War II: "Looking out my window from the office I can see the crowds gathered along the sides of the street where Eisenhower is expected to travel from

the Farm Field Day to the Des Moines Air Port. I wish I could expect
that all that crowd—and there are plenty of them—were going to vote
the Republican ticket this fall.''[17]

As the 1956 campaign heated up, Darling received word of an
organization calling itself "Conservationists for Stevenson-Kefauver."
He noted that the names he recognized were those of out-and-out ad-
vocates and promoters of government-owned and controlled hydroelectric power plants.[18]

Following the election and Eisenhower's victory Darling was advised that, with an internal reorganization of the Fish and Wildlife
Service, his twenty-two-year-old appointment as a deputy game warden
would be continued. Darling was surprised to learn that the appointment had not been terminated years before. The eighty-year-old Ding
happily accepted the continuation noting, "There have been a number
of occasions on which I wished I had the authority of a Deputy
Warden. . . ."[19]

Darling was favorably impressed by Fred Seaton's performance as
interior secretary. "He is a good scout and a good friend of conservation
when the Brass Hats don't interfere," he wrote. With some help from
Clark Salyer, Darling also became a believer in Ross Leffler, Seaton's
assistant secretary. He was even encouraged by actions of Hubert
Humphrey, the liberal Democrat from Minnesota who was trying to
save some of the nation's wilderness areas. "I guess we have been
waiting for you," he wrote, "to weld into practical form the legislative
measures which if enacted would carry out the dream of millions of conservationists in this country."[20]

Ding admired the espousal and practice of conservation wherever
he found them, and he perceived a pronounced streak of both in those
of the Mormon faith. He unveiled this observation in a letter to his
friend Clarence Cottam, who had become head of the Welder Wildlife
Foundation in Sinton, Texas:

I have often thought and wondered about the impulse and training you received from your Mormon heritage and whether that society didn't have
something that the rest of us have missed by several rows of apple trees. You are
not the only one of Mormon extraction who has stimulated this same curiosity
in me. I am sure it is no accident that all of them have had some degree of the
devotion and mental discipline of which you are a conspicuous example.[21]

His own mental discipline kept Darling on the lookout for
developments in the red tide controversy. In 1958 he read in *Scientific
American* about the causes of poisonous tides. He wrote to Carl Miner

that one of the twelve organisms illustrated in the article "most nearly resembles the organism which we found so prevalent in the yellow waters off the beach of Captiva." Darling took exception to the journal's illustrations: "I am surprised, though, that the sketches accompanying the article showed no cilia, which under high magn[i]fication appeared covering the body of the organism and which, I am sure, gave the body the swirling motion rather than the flagella, as indicated in the article."[22]

He had earlier studied an article on the "History of Red Tides and the Dinoflagellates" in the London *Illustrated News* and brought it to the attention of the Marine Laboratory at the University of Miami. In 1954 the laboratory sent Darling a copy of its complete report on the red tide. Darling wrote to Miner, "It is strange how long it has taken the so-called scientific agencies in charge of wildlife and natural resources to recognize the seriousness of that infestation which began, for us, in the winter of 1946–47."[23]

Darling's skeptical attitude toward learning and publishing grew almost cynical. He scolded intelligent conservationists who had something to say, who said it well, who wrote books, and who then considered their work done. He recounted his efforts aimed at teaching teachers to teach conservation and the failure of several approaches to bear any fruit. The writing of a textbook for the teaching of conservation, Darling wrote, "has been tried so many times and so many unsuccessful textbooks on the teaching of conservation are in print without being used in places where most needed that I am filled with misgivings over another attempt."[24]

The former chief also took some pride in having taken charge of a Biological Survey that was populated by a large number of persons with Ph.D. degrees in various fields of specialization and in having stood the agency back on its feet. When Fred Lincoln, a colleague from Ding's Survey days, received an honorary doctorate, Darling wrote a letter of congratulations in which he noted, "I may be a little confused on this 'Doctor' business but when I was first introduced to the staff of the Biological Survey it seemed to me that I was completely surrounded by doctors' degrees. Everyone seemed to be a Doctor except me!"[25]

In the midst of putting his affairs in order, Darling had also decided against investing any of the residual value of his estate in scholarships for college and university students. He had done an informal study, he claimed, of students with scholarships "as compared to the standings of those who had no scholarships but sawed wood, tended furnaces, washed dishes and waited on tables" to pay their tuition. Darling's findings indicated that "the scholarship men" were far below the normal average. His observations were gleaned, he said, from his terms as a trustee of "a couple of small colleges." Even worse, Darling complained, scholarship winners were an ungrateful lot and likely to spoil. He reported:

Alas two of them, and one of them a girl from a neighboring town who came highly recommended to me, had learned through the ease with which she financed her college education that it wasn't necessary to work for a living and the last I heard she had become a professional panhandler and her husband, a former WPA boondoggler, had the nerve to solicit me for a donation for the rescue of fallen stenographers out in Seattle, Washington—a good cause, maybe, but it was a fake as I discovered after some investigation.[26]

Darling's negative attitude extended as well to his efforts to halt the Army Engineers in their dam march to the sea. He was critical of Governor Les Miller of Wyoming and Herbert Hoover and of conservation organizations that would not cooperate:

It's a sorry day when men of the caliber of Les Miller and Herbert Hoover will sell the water resources down the river to the Army Engineers as the only cure they know of to defeat socialization of water power. We are licked again and this time down for the count and the tragedy is a little harder for me to bear because this is the kind of fight we might have won had the [National Wildlife] Federation functioned as designed.[27]

Darling wrote to Clark Salyer, "I finally severed the umbilical cord and no longer even recognize the Federation as a legitimate offspring!" He added, "That metaphor is somewhat mixed but so are the chromosomes and genes in the Federation. I was supposed to be the father of the Wildlife Federation but Carl Shoemaker and his cronies and aides took turns sleeping with the bride before the honeymoon was over and I now look upon the bastard as beyond redemption."[28] To Shoemaker himself, Ding wrote that the organization had forfeited its right to the title of Federation:

I confess that I blush with shame every time faithful conservationists greet me with a smile of confidence and tell me that every year they contribute their dollar for wildlife stamps in support of "my conservation Federation." My God, what a magnificent deception I have allowed to exist. We have no more federation than a bunch of wild rabbits. . . . I had hoped that the Federation would be the crowning achievement of my devotion to Conservation. It is, instead, my greatest humiliation. Yes we have no Federation![29]

Darling's defeatism became a reflex. While he continued to condemn the activities of the Wildlife Federation,[30] he also issued a tirade against Wildlife Preserves Incorporated and dismissed it as just another conservation organization without benefit of consolidation:

I can think of nothing in the conservation field which promises such beneficial results as a unification—you might call it a "Clearing House"—of the all too numerous federations, societies, chapters, clubs and services, each maintaining a staff of paid executives, each owning or renting office headquarters and each spending enormous sums annually in publications for general distribution, and all saying the same thing in just so many words.[31]

Darling later wrote a letter of abject apology for his ill-considered criticism. He also subsequently spotted a ray of hope for the National Wildlife Federation, with the appointment of a new staff member; but he doubted if any man in the world could regain the lost ground which the Federation had held in the beginning.[32]

Darling's biting criticism, devoid of the characteristic Ding humor, marked his correspondence in the mid-1950s. No doubt his own declining health and the sudden and shocking deterioration of Dr. John's physical and mental health contributed to the conservationist's mood and his pervasive negativism.

In 1954, virtually at the completion of a yacht trip up the Mississippi River from Florida, Darling was hospitalized in Saint Paul, Minnesota, with a heart condition and was released after approximately five weeks "with a large vacuum in my wallet." Because of his history of lung ailments, Darling was especially vulnerable to pneumonia. He wrote, "Had it three times and have now exhausted all the potentials of the anti-biotics and the medics warn me that I probably won't survive another attack." He tired quickly and, even though his mind raced through issues, questions, observations, and ideas that might have been helpful in the conservation wars, physical exhaustion forced him to cut short the dictation of letters to colleagues. He wrote of his drained energies, "I'm too old to be expected to do anything . . . and with the increasing years the instinct to fight for the many lost conservation causes has dribbled away until there is almost no fight left in me."[33]

As he hovered near the eighty-year mark, Darling acknowledged that the doctors were "doing their damnedest" to keep his heart going. His despondence deepened. "I don't even worry about it and if the affairs of state keep getting into more trouble it won't make me very sad to go around the bend and forget it all," he revealed. His spirits later improved. He claimed he had a "couple of good fights left in me," and he wistfully wished he "were young enough to take up arms and get back into the battle." Within another year, however, he admitted "a tremendous fatigue—maybe disgust is a better name for it."[34]

Darling felt lonely and isolated from the major conservation issues of the day. He described one correspondent as being "wonderfully good to me" for writing to him occasionally. He still controlled a spark of life, he wrote Clarence Cottam, "if anybody would listen to me, which they don't."[35]

As Darling tried to make his way through the discouragement and depression that darkened his world, one of the brightest lights in his life was unexpectedly extinguished. His son John, the promising physician, had practiced at several medical installations and in 1956 moved with his wife and two daughters to Oak Park, Illinois, near Chicago. Disaster struck at John's nervous system. By early 1957 Dr. John's condition was described by his father as improved, although the young physician was suffering periods of retrogression and of "low depressive spells" which were growing shorter as "rigid management and relief from strain are provided." Dr. John was later moved to Tarpon Springs and Jay and Penny stayed at Captiva much later than usual in 1957 so they could make weekend visits to their son and could help keep his "spirits from languishing." By September of that year, Ding had resigned himself to the prospect that "John is likely to be a patient for the rest of his life at Anclote Manor, near Tarpon Springs. . . ." He and Penny decided to give up their winter quarters on Captiva in favor of living arrangements nearer their son.[36]

Dr. John's career as a surgeon had been destroyed by epilepsy. His unpredictable seizures were traced to brain damage suffered when the car he was driving left the road and struck the bridge near New Hampton, Iowa, that wet, chilly night nearly twenty years earlier. The same malady eventually led to unpredictable and irrational behavior that was a constant source of distress and strain for his father. Penny Darling, a modest, tiny woman, rose to the occasion. A graduate of the University of Wisconsin, Penny had taught crafts to epileptics. As John's condition continually worsened, Penny was far more capable of handling the situation than was Ding.

In the fall of 1957 Darling again entered the hospital for a complete rest and was restricted to an oxygen tent for an entire week. He suffered from insomnia and complained, "I'm just an old dried husk of what I used to be."[37]

Also in the mid-1950s Darling had been obliged to relinquish two homes that held symbolic significance for him, his family, and his admirers. His beloved Fish House off the coast of Captiva became too burdensome to maintain and was too far from Tarpon Springs to be efficiently useful. After he sold the unique structure, he wrote, "I too had reverence for the last resort of free men—The Fish House—and I am sick to my stomach every time I remember that I no longer possess it . . . when I found someone who wanted the Fish House more than I did I let it go.[38]

For several years, Darling had also tried to sell his magnificent but cumbersome Des Moines home at 2320 Terrace Road, immediately south of Terrace Hill, the old Hubbel mansion.[39] Jay and Penny had purchased the core of the home, but their several additions to the structure gave it a distinctive character and charm. Darling—whose artifacts, furniture, and books had been accumulated over a lifetime from a large part of the world—fretted about what to do with his belongings if he and Penny were to sell the home and take an apartment. On one side of the house Jay had added a huge living room of his own design. The open, two-story chamber featured huge, hand-hewn wooden beams, leaded glass windows, and a fireplace. Along one wall he built a wooden-railed catwalk leading to a windowed niche in one corner of the room. He spent many hours working in his "Monk's Corner." Darling the musician occasionally sat alone in the large room, playing his accordion and awaiting ideas that had failed to float up to the Monk's Corner.

Darling's anxieties about the house were erased when a young couple, Mr. and Mrs. Ed Hunter of Des Moines, agreed to purchase the home as it stood. The Hunters, devoted Ding fans, had three boys and were expecting their fourth child when they purchased the house in 1955 "with the mouthwash in the medicine cabinet and the flour in the cannister." Darling was delighted and so were the Hunters.

Also in 1955 Jay and Penny joined Forest and Fae Huttenlocher on a world tour, the major purpose of which was to give Darling the complete rest his doctors insisted upon. Huttenlocher was a Des Moines insurance executive and good friend of Ding's. Fae Huttenlocher was an editor at *Better Homes and Gardens* magazine. The foursome boarded a Norwegian cargo liner and for $1,150 per person, including meals, toured the world for approximately four months.[40]

Darling, who would not allow the trip to be restful for anybody aboard, shed his dour disposition and once again displayed his bright knack for play. "He saw the funny side of everything," Fae Huttenlocher recalled. "We laughed all the way over there." Mrs. Huttenlocher's employer, Ed Meredith, had urged her to bring back an exotic animal from her world tour. Meredith wanted the animal for a business promotion and planned to donate it later to a children's zoo, then in the planning stages in Des Moines. At Darling's suggestion, she obtained a cobra in Singapore, complete with a glass-sided suitcase and a small horn. Darling delighted in entertaining guests of the ship's cap-

tain by "doodling on the horn and making the cobra do its dance."
The cobra never made it back to the United States, however. Its fangs,
once removed, were growing back; its stench was growing tiresome; it
had taken over the parlor adjacent to the Huttenlochers' cabin; and it
required a large daily diet of frogs. The snake was left in Ceylon.

In 1957 the Darlings traveled to Trinidad and Tobago on an Alcoa
ore boat. In 1958, at about the time the islands of Quemoy and Matsu
flared into international headlines, the Darlings traveled to Japan,
where Ding was amused by the scale of duck hunting. "There were a lot
more hunters than there were ducks to be sure," he reported.[41]

The alert Darling made efficient use of his travels, casting his
observations against a background of concern for conservation of national and international resources. His spirits were buoyed by his trips,
but the effects were only temporary. He felt cut off, stranded on an ox-
bow, shunted out of the mainstream of national conservation events.
He was more than eighty years old, his contemporaries were falling by
the wayside, and his health denied him the freedom to jab his op-
ponents and to cleverly sidestep their thrusts. When he did muster the
strength to swing at an adversary, he was sometimes ignored. Even
though it was politely done, it was painful to an octagenarian who had
devoted most of his generous allotment of years to saving something for
the future.

# 20

## *Fight to the Finish*

JAY DARLING'S HEALTH was bad and growing worse, and as the turn of the decade from the 1950s to the 1960s approached it seemed to him the vital signs for conservation in the United States were as uncertain as his own. On the state scene, the Iowa Conservation Commission became embroiled in its most bitter and bizarre conflict since Darling had aided in its creation. At the national level, he was frustrated by the success of the dam builders and the Florida developers who seemed to plunder the best of what was left of the nation's natural resources. He was incensed at what he believed to be past malfeasance in the administration of the Department of the Interior, which had stood helpless or, even worse, had lent a hand to the looting of the nation's bounty; and he was worried about its happening again. The view was bleak in several directions.

Darling was encouraged when Ira Gabrielson was employed to conduct a study of Iowa conservation policies and practices. When Gabe's work was completed, Darling conceded that the report was helpful; but he took exception to what he called "Gabrielson's complete whitewash of the Governor Blue administration and its pernicious interference in the operations and program of the State Conservation Commission and staff." Darling observed that the stature of commission members had declined and that a "round of knuckle-headed governors" who wanted to control the money had emasculated the board's effectiveness. Darling lauded the efforts of Bruce Stiles, director of the Conservation Commission, who was obliged to carry the burden of the fight to keep the commission inviolate.[1]

In July 1959 Darling wrote an illustrated letter, featuring an agitated milk cow, to Iowa Governor Herschel C. Loveless:

146

I am sure you know what happens when a new hired man tries to come up on the wrong side to milk a cow. It doesn't take long for the word to get around that the new hired man doesn't know much about his business.

Believe it or not, that is what experienced and well informed conservationists are saying about the "about-face" and "to-the-rear-march" policies of the newly appointed majority on the State Conservation Commission.

Darling contended that a district method of administering conservation practices in the state had been tried in Iowa and elsewhere and had failed. He argued for statewide policies by a commission devoted to the entire state and devoted as well to the application of scientific principles to conservation practices. "It seems a shame to throw that method out the window," Darling wrote, "and become the laughing stock of every conservation agency in the country."[2]

The new commission also voted Stiles out of his job and, according to Darling, two commission members visited the hospitalized Stiles to request his resignation. According to a press release issued by the commission, Stiles had been released from his duties due to his poor health; but a commission member informed Darling that it was a dismissal.[3] Darling thought the actions of the Conservation Commission bordered on the barbaric. He related the details:

Bruce Stiles . . . has been Director for eight or ten years at about half the salary of the Directors in the states surrounding Iowa. He had courage and enough technical knowledge to appreciate the value of good technical men in the Fish, Game, Water and Soil Management Divisions. Bruce went to the hospital for a physical checkup the week before the new Commission had its first meeting.

They found a malignant tumor in the chest of Bruce Stiles and the news of it was published the morning of the first meeting of the new Commission. The Commission voted Bruce Stiles out of office and two of the new commissioners went to the hospital late that afternoon to demand his resignation or be fired. I can think of no more brutal thing for a Commission to do. Bruce died two days later after an operation.[4]

In the midst of the gloom there was a hopeful sign. One of the new commissioners was a Des Moines man, an insurance executive and bank board member by the name of Sherry Fisher. Soon after Fisher's appointment Darling predicted that the new appointee was going to be a fair-minded and intelligent member of the commission. Darling's first impression grew into a lasting one. He later remarked, "On the whole the Commission is now doing pretty well. The new members are pretty lively chaps and Sherry Fisher, for instance, of Des Moines, seems to me one of the livest members we've ever had on the Commission."[5]

As Darling's favorable impression of Fisher and the commission grew, his opinion of Stiles became tarnished. As time passed, Darling conceded that he had overreacted at the time Stiles was asked to resign. Ding learned that Stiles was ''not very well liked and didn't get along well with the legislature'' and that the governor and many legislators felt Stiles was not providing adequate service.[6]

I found that both Bruce and to some extent the members of the Commission had been derelict in some of their responsibilities. Bruce had been in ill health for some time and had been drinking more than he should. Members of the Commission who valued the excellent qualifications which Bruce had demonstrated over and over again were what you might call overly loyal to Bruce, and allowed conditions to drift along without investigation.

Darling put it metaphorically, ''I found that the kettle had been boiling over all over the stove and everybody had been so busy wiping up the mess to save Bruce Stiles that there really wasn't a pair of clean hands in the whole outfit.''[7]

Darling even had a good word for Governor Loveless: ''The Governor has played quite a straight game all the way through. In fact I wish we had a few Republicans who looked as well under the X-Ray machine as Loveless.'' Concerning Stiles, Darling concluded, ''the best thing we can do is to remember the years during which he gave splendid service to the state and the State Conservation Commission. . . . I keep those stories [of Stiles' later years] to myself, and the frailties which he showed in the last few years I hope will remain unknown to the great fraternity with whom he stood very high.''[8]

Darling remained on guard, nevertheless, for ruptures in the wall of political protection he had thrown up around the Conservation Commission many years earlier. He noted that for forty years prior to 1931 ''every Governor of Iowa dictated his own private conservation policy and our game, fish, waters and soils went down, down, down the rat hole to an approximate zero.'' The first line of defense, Darling maintained, was a Conservation Commission made up of intelligent lay persons with the good sense to leave professional decisions to the professionals. He often relied upon one of his medical metaphors to make the point: ''It was not intended that the civilian members of the Commission would presume to take over the scientific application of the biological management any more than that the civilian trustees of a hospital would be entitled to perform surgery in the operating rooms.''[9]

Darling also argued that if the commission was to attract a professional staff it would have to provide salaries more nearly commensurate with the professional requirements. In particular, he believed the salary for the director of the Conservation Commission should be raised considerably and that it was the commission's responsibility to obtain the necessary funds from the legislature. Too many commissioners, Darling asserted, who thought they knew as much as the experts, were willing to serve for nothing.[10]

As the 1960 elections approached, Darling was as optimistic as he had been in years about Iowa's progress in conservation. "On every hand," he wrote the director, Glen Powers, "I have been hearing that we have the best Conservation Commission and the best organization of technicians and director that Iowa has ever had!" The euphoria did not last long however. Soon after the elections Darling wrote Iowa Governor-Elect Norman A. Erbe, concerning the appointments to the Conservation Commission: "I'm not going to ask you to appoint any particular person but I just hope you'll be careful not to fill the vacancies with dummies."[11]

Following the announcement of Erbe's appointments, Darling let the new governor know of his discontent:

I would like to remind you that it was the Republican Party, Republican Governor and Republican Legislature who in 1930 set up the original Conservation Commission and voted the 25-Year Program. Those folks aren't all dead yet and they remember, even if you don't.

Three Governors in a row making such stupid appointments are enough to wreck the effectiveness of the Commission.[12]

A flurry of correspondence ensued. Erbe argued that he had consulted conservationists for suggestions of names for commission members, and Darling insisted that he knew names of qualified persons urged upon the governor but not appointed. Darling accused Erbe of "persistent double-crossing of everything in the conservation category. . . ."[13]

Long before the 1960 elections Darling was uneasy about the condition of the people and the direction of the policies in the Department of the Interior. His recruit to the conservation wars, Clark Salyer, had suffered a heart attack, was blind in one eye, and experienced occasional blindness in the other; but he maintained a rigorous work schedule in the defense of the refuge system that he and Ding had created and that he oversaw. While Darling worried about Salyer's health, he worried as well about who would replace Secretary Fred Seaton and Assistant Secretary Ross Leffler following the elections. Darling had been pleased with the conservation principles and courage exhibited by Seaton and Leffler, and he suspected that, whoever won the presidency in 1960, there would be a change in the leadership of the Interior Department.[14]

As Richard M. Nixon campaigned for the Republican nomination for president, Darling was struck by some of the vice-president's observations on population. He contrasted Nixon's views with those of President Eisenhower: "There is a great improvement in your attitude toward explosive populations and their control over those expressed by President Eisenhower. I hope the time will come when you may speak out even more forcefully on the subject. I regret that we must conclude that Eisenhower and all graduates from West Point, particularly the members of the Engineers Corps, are biological illiterates."[15]

He was earlier chagrined, however, at Nixon's failure to respond to a personal note:

Who in hell sorts your mail? I get a personal letter from you thanking me for my contribution to the Republican Campaign, but completely ignoring a brief personal letter from me to you.

The letter I wrote you concerns things a damned sight more important to the country and to the success of the Republican Party than any financial aid that I can give. What the hell!

Darling also expressed his concern about the future leadership of the Department of the Interior to Nixon and urged him to "consider the continuance in office of Secretary Seaton and Assistant Secretary Leffler."[16]

Ding watched the national political conventions on television that summer, but he found little there to cheer him up:

I have just been writing Fred Seaton that I'd been seeing him at odd intervals, via the T-V, at the Chicago Convention, but that was the nearest approach to the thought of Conservation of Natural Resources that I'd seen at either Convention, and how did he explain the total eclipse of such an important subject when the political leaders and delegates from every corner of the U.S.A. assemble to consider the future welfare of the U.S.A. Kinda discouraging, eh?[17]

He commiserated with his friend Clarence Cottam:

I suppose you may have listened in on the national conventions in Los Angeles and Chicago, but one thing I'll guarantee that you did not hear and that, particularly, was anything about the conservation of natural resources. I don't see how in hell so many representatives and leaders in the field of national interests could tear the English language to tatters for two weeks and never mention the subject of natural resources.[18]

Darling's involvement in state and national conservation issues seemed to charge him with new vigor and enthusiasm. By September 1960 he wrote that he was "too busy to even stop and get my hair cut."

The following month, he promised a fellow waterfowl protectionist, "I'll keep the wires hot as long as I live."[19]

Following the elections Darling lost little time wiring President-Elect John F. Kennedy:

In consideration of the new Secretary of Interior and Assistant Secretary in charge of Fish and Wildlife, I urge special attention to the precarious state of the migratory waterfowl population, the conservation of wetlands and the wilderness areas, all of which require men of good technical qualifications. Partisan political appointees are not very good at it.[20]

When Kennedy appointed Stewart L. Udall to head up the Interior Department, Darling wrote the new secretary,

God only knows with what anxiety the old torchbearers in the field of conservation have watched for this appointment, and we are all praying that your Assistant Secretary in charge of fish and wildlife resources, on whom will depend the salvage of our wetlands and the population of migratory waterfowl, will be the kind of chap who will take good care of these very important resources.

Darling had a candidate in mind for the job. When Udall replied to Darling's letter, Ding wrote again urging the secretary to carefully examine the credentials of Clarence Cottam. "He would be the best man I know for Assistant Secretary of the Interior," Darling wrote:[21]

The appointment of Udall was an encouraging sign to Darling. Although Ding admitted he knew little about the man, he did know that Udall was a member of the Church of the Latter Day Saints. "I had six or seven Mormons in the Bureau of Biological Survey during my brief administration," he recalled, "and those Mormons were the best men on my staff." Cottam, also a Mormon, was a Republican and chose not to declare himself a Democrat to enhance his appointment chances. "I can't conceive of anyone hankering for the job if he wants to do a creditable piece of work," wrote Cottam. "The poor devil who takes that assignment as Assistant Secretary, if he does his duty, is . . . sure . . . to be as popular as a polecat in Sunday School. . . ." Frank P. Briggs, a sixty-seven-year-old publisher and former U.S. Senator from Macon, Missouri, was appointed assistant secretary February 6, 1961.[22]

Darling had hardly settled back when a new crisis loomed in Washington, D.C. A bill to establish a federal wilderness preserve was being held up in the Senate Interior Committee, and Darling could not understand why. He wrote Senator Clinton P. Anderson, the bill's sponsor:

It is thirty-six years ago since top scientists and land and water conservationists worked on the details of just such a measure. Those of us who are still alive are very grateful for your embodiment of the fundamental principles in a measure which should meet the approval of the U.S. Senate. What reason has anyone to try to trip up and strangle this measure which everyone in the country is for except a few draggle-tails in the Interior Committee?[23]

He also wrote to senators Barry Goldwater, Quentin N. Burdick, Lee Metcalf, Frank Church, Henry Dworshak, and Ernest Gruening. Iowa's Senator Jack Miller was also a member of the Interior Committee and Darling joined in sending a shower of telegrams to Miller Wednesday, July 12, the day before the committee vote on sending the Wilderness Bill to the Senate floor. Darling's own telegram read: "A lot of your Iowa friends are going to be very unhappy if they find you holding back the Wilderness Bill from the floor of the Senate."[24]

Darling wrote Iowa's senior senator, Bourke B. Hickenlooper:

Speaking of the Wilderness Bill in particular, there never has been a time in the last fifteen years when the cattle and sheep grazers association, the landgrabbers and lumber hijackers, and the mining interests, couldn't dig up controversial issues to defeat any wilderness bill that could be written. I am sure that Jack Miller, hearing their arguments for the first time probably, was greatly impressed, but those who have fought those same interests beginning with the Taylor Grazing Act know those controversial issues by heart.[25]

After the committee discussed a damaging amendment the following morning, Miller voted for it. When the amendment was brought up for reconsideration in the afternoon, however, Miller reversed his position in a 9–8 vote and helped pass the bill to the full Senate by a wider margin, despite heavy committee representation from western cattle and mining interests. Senator Anderson reported the results to Darling and suggested that a *Register* editorial in Miller's behalf would be "appreciated by me." Darling replied to Anderson, "We strive to please," and enclosed a clipping of an editorial from the July 16 issue of the Des Moines *Sunday Register.* The editorial, titled "The Wilderness Bill," read in part, "Many Iowans will find satisfaction in the fact that this state's member on the committee, Senator Jack Miller, voted for the bill."[26]

While Darling found some gratification in the progress of the Wilderness Bill, he saw one defeat coming after another on the nation's waterways. There seemed to be no stopping the huge dams and, to add insult to Darling's defeats, the Ding Darling chapter of the Izaak Walton League came out in public support of the Red Rock and Saylorville dams in Iowa.[27] Darling was unusually reserved and indirect in his response to the Darling chapter:

I have tried for a long time, without much success, to think of something I could do to merit the honor you have done me. Today when I received a copy of your Resolutions endorsing the Red Rock and Saylorville Dams the percussion jarred loose an idea and I am enclosing $100 as a first installment of what I hope will prove an annual project as long as I live and maybe longer. You may use the $100 in any way you choose but my preference would be that it go toward paying the expenses of a qualified delegate from your Chapter to attend a convention, either State or National, dedicated to the best scientific knowledge and methods available for enlightened conservation and the avoidance of high pressured policies which may prove more destructive than helpful.[28]

Darling wrote to the executive secretary of the Iowa division of the Izaak Walton League of America:

I have no pride in my opinions and I have been wrong as many times as I have been right but this epidemic of Big Dams up and down the main waterways of the middlewest are most of them pretty bad. They will be so full of mud after the first twenty years of their existence that they will be worthless either as flood control reservoirs or hydroelectric power plants, and that chain of dams up and down the Missouri River are going to submerge under impounded water the best of the tillable land throughout the Dakotas, and all along the Missouri Flyway. If anyone wants to see how those dams influence recreation, wildlife and particularly fishing, all they have to do is take a trip down to Keokuk [Iowa] and examine the river above that dam, for thirty five miles. It is practically a biological desert.[29]

The following month, a banner headline in the Des Moines *Register* might as well have been bordered in black for Darling's reading. It read "D.M. Area Dam Funds Voted," and the article beneath the headline reported that the U.S. Senate had appropriated partial funding for the $71-million Red Rock project and the $47-million Saylorville Dam. Darling was stunned, but he was still on his feet after years of losing to the dam builders. He saw hope of saving the Missouri River, on whose banks he had matured. He rummaged through his mind and dusted off an idea that had lurked there since being packed carefully away a quarter-century earlier. On one of his weekend visits with Herbert Hoover at Rapidan Camp, Darling had suggested to the president that he forget plans to create a nine-foot channel down the Mississippi River. "My argument was that he had better leave the old Mississippi alone," Darling wrote, "'and declare the waters and its broad valley a national park, from Lake Itasca in Minnesota to New Orleans.''[30] The idea could be as easily applied to the Missouri River, Darling reasoned:

I have lived along that Missouri River since the middle-eighties, when my family moved to Sioux City. I have hunted and fished and camped and swam in those muddy currents for just about a half a century [presumably a calculated exaggeration] before I moved over to Des Moines, and I know the tremendous potentialities as a game and fishing area if it wasn't for the invasion of the Army Engineer Corps, land speculators and quarrels over jurisdiction between the Iowa and Missouri state agencies. My suggestion was that the two states get together and declare that Missouri Basin from the west bank on the Nebraska side to the east bank on the Iowa side an interstate park. It would make a grand one and I'd a darned sight rather have the Missouri River for Chief Engineer than I would Army Engineers.[31]

Darling contended that the meandering river would continually cause border disputes and damage to farmland, communities, and homes. Flooding would occur only infrequently, he argued; and when it did, it would cover only federally owned land that was being preserved for the benefit of all. He noted that millions of dollars had been invested in control of the Missouri River ''and still the river disregards all the plans of the Army Engineers. . . .''[32]

Darling's uneasiness about the future of the Missouri River Valley led him to the expression of another related idea. In May 1961 he called his protégé on the Iowa Conservation Commission, Sherry Fisher, and asked him to a meeting in his office in the Register and Tribune Building. (Executives of the newspaper firm made Darling's office available to him after he retired and insisted that he office there the rest of his life. But, Darling wrote: ''I don't think they calculated when they made that generous statement twelve or thirteen years ago that I was going to live so doggone long.'')[33] Darling told Fisher he wanted to incorporate the Missouri River into a national outdoor recreation and natural resources ribbon along the historic trail of Lewis and Clark. Fisher noted that Darling's proposal came at a time when the Missouri was largely uncontrolled, with no wing dams and no cutoffs or oxbows. The area, in addition, would create what Darling called ''an avenue for wildlife.'' Although Darling's health was poor, Fisher recalled he was bubbling with excitement over the prospect. He looked Fisher in the eye. ''I can't live to do these things, but I'd like to know you'd try to do it for me,'' he said. ''I'll try,'' Fisher promised.

The aging conservationist had other irons in the fire. He continued to encourage meetings of conservationists and editors of the most influential newspapers and magazines. Relations with the editorial chiefs at the *Register,* however, had become strained. Darling claimed that the editorial writers and he had not been on speaking terms since W. W. (Bill) Waymack left his positon as editor-in-chief to become a member of the U.S. Atomic Energy Commission. "For a long time," wrote Darling, "the editorial writers claimed that I, a pronounced Republican conservative with front-page cartoons, undermined their influence as left-wing progressives. I hope to hell I did to some extent!" Darling had kind words for Ken MacDonald, however: "Ken Mac-Donald was formerly Managing Editor and his authority stopped short of the Editorial Page. Now he is in command and I recommend him highly."[34]

Darling suggested that Bernard Baruch, or someone of equal prominence, be asked to invite the publisher of a major metropolitan daily newspaper to dinner, preferably in New York City and at a comfortable time of the year. Other guests would include several articulate conservationists who would give the publisher "the works." "The works" would constitute a "mass innoculation [*sic*] of conservation virus" for the publisher. Darling even had a publisher in mind—John Hay Whitney, new head of the New York *Herald Tribune.* Darling had difficulty understanding the indifference of the popular press toward conservation: "It seems to me amazing that none of the master minds of the editorial press have ever considered this cause as one of their responsibilities as a major agent in adult education."[35]

The related idea of a clearing house for conservation information received fitful attention from time to time but never seemed to take flight. Darling's friend Arthur Carhart initiated efforts to create a conservation collection at the Denver Public Library, and Darling saluted it. The conservation collection, however, was not the answer to a conservation clearing house. In response to a plea for funds for a conservation-related organization, Darling noted, "there are so darned many conservation organizations that I can't possibly give to all of them." There remained a need for a consortium of such organizations, said Ding, for the sake of efficiency and effectiveness: "It was that objective that stimulated the formation of the National Wildlife Federation, which has now lost sight of its target entirely, and I could never understand why the need of such a movement is never mentioned any more."[36]

Darling, meanwhile, was honored by *Nature* magazine as one of a handful of changemakers in the world of natural resources management and preservation. He was selected for outlining a system through which all concerned organizations could unite their efforts in the interest of conservation.[37] Darling was embarrassed:

I blush to think of the optimistic prophecies and promises which I held up before the eyes of the hopeful public which, alas, never came true. My face is reddest when I think of the Wildlife Federation and the way it went astray and finally flopped, as shamefully as a thousand-dollar Llewellin setter, which forgot all about the Quail cover and went bounding out over the hills and dales in pursuit of a jackrabbit.[38]

Nor, as the years wore on, had Darling forgotten his interests in Florida, which "comes nearer being blind in both eyes to its wildlife and fish endowment than any state in the union. . . ." Darling no longer stayed at Captiva. He and Penny spent their winters at the Harrison Hotel in Clearwater, where Dr. John and his family resided, close to Tarpon Springs on Florida's west coast north of Fort Myers. Even so, he took interest and a hand in objecting to plans to construct a causeway from the mainland to Sanibel Island. He was also heartsick when he learned that Sanibel and the surrounding area had been struck by a hurricane, and he sent a contribution to help clear the destruction. The protector of waterfowl was disheartened by reports that the trees in the mangrove swamps and on the tiny islands were leafless but festooned with dead American egrets, snowies, and greater white herons. The trees were also nearly flattened on the tiny key set aside as a refuge for the Key deer that Darling had battled to protect.[39]

Between bouts with serious illness, Darling kept up his pace and his interests, determined to surmount whatever barricades were put in his path. In 1959, as he lay critically ill, it became apparent to him that he might not survive. Lying in bed, he sketched a "farewell" cartoon, a likeness of his office at the *Register* with a phantom figure of himself rushing out the doorway. The cartoon, titled " 'Bye Now—It's Been Wonderful Knowing You,'' he gave to Merle Strasser. In the event of his death, Ding instructed her to give it to Ken MacDonald. Merle was relieved when Darling returned from the hospital and the cartoon remained in her files.[40]

In 1960 Darling began work on what was intended to be a souvenir booklet of Ding cartoons for private distribution. The "booklet" grew into a cloth-bound book titled *It Seems like Only Yesterday*. The copies for Darling's personal use were printed at his own expense. His friend in conservation and boyhood, Max McGraw, who became president of the North American Wildlife Foundation, had many more copies printed and distributed them widely to friends and colleagues.[41] Other collections had been published previously. Darling, with the assistance of John M. Henry, had earlier assembled cartoons for a special book, *As Ding Saw Hoover*. In 1960 the Pioneer Hi-Bred Corn Company, headed by James W. Wallace, the brother of Henry A. Wallace, published a collection of Ding's work in a book, *Midwest Farming as Portrayed by a Selection from Ding's Cartoons.* Donald R. Murphy, a member of the *Wallace's Farmer* editorial staff during much of Ding's cartooning career, edited the book.

Following the turn of the decade, Darling was showered with recognition from many quarters; and in the warmth of the approbation, his grim disposition seemed to soften and his humor reappeared. A beautiful new variety of flowering crabapple tree in Des Moines' Denman Park was named the Jay Darling Crab. "Around here they call it the 'Ding Crab' and I'm flattered to death because it is the most gorgeous tree when in flower, and also when in fruit, that I've ever seen," Darling wrote. He was named to the national Izaak Walton League of America Hall of Fame. He was named the winner of the prestigious Audubon Medal for service to conservation, and he received The Garden Club of America Florence K. Hutchinson Award. The Des Moines Press and Radio Club set aside an evening to pay homage to the elderly Darling. He was made an honorary member of Teddy Roosevelt's Boone and Crockett Club.[42]

The warrior for wildlife had other reasons for satisfaction. The hatch of waterfowl was up in 1960, and Darling reported joyfully, "The season doesn't open for another week but the marshes and lakes are already filling up with Mallards, Teal and Gadwalls, and it is much easier to make a count now than after the shooting season opens." He learned that Canada goose production on the Lower Souris Refuge was on the upswing. A report noted that ninety-five active nests had been found, thirty of which were on Ding Island alone. Observers wrote, "Named for him, we suspect our old friend 'Ding' Darling would enjoy knowing this." He learned as well of a quiet move afoot to fund a crash program for restoration of national wetlands. "That was the most exciting bit of news I've had since eighteen hundred and God knows when!" he exclaimed. He finally viewed the activities and leadership of the Iowa Conservation Commission with satisfaction: "We have, for the first time, a really active, working Conservation Commission, and a new Director, Glen Powers, and they have a better program and are working harder on it than anyone we've had in that department for twenty years."[43]

For the first time in many years he also perceived an interest in conservation on the part of a nationally circulated popular magazine. He wrote the publisher of *Sports Illustrated* to congratulate him on the magazine's interest in the conservation of natural resources. His memories of the Biological Survey, with the perspective afforded by many years, also mellowed. "Looking back," he wrote, "I think the approximately two years I spent with the old Biological Survey were the two most exciting years of my life. . . ." He wrote kindly of Henry Wallace and the former vice-president's role in the New Deal: "As a result of that experiment in Iowa [Survey and 25-Year Plan], the State of Missouri followed, but not quite so efficiently, and then in 1934, '35 and '36 Henry Wallace, who had watched what had been happening to his own state, sought to apply the same principles through the efforts of the Federal Government and emergency jobs for the unemployed." Darling also suggested that there "were better men than I leading forces battling for land and water management, Hugh Bennett and Henry Wallace for instance."[44]

Finally, he even made peace with the National Wildlife Federation when in 1961 he agreed to serve with his friend Walt Disney as cochairman of the 1962 National Wildlife Week, sponsored by the Federation and scheduled March 18 through 24. Darling wrote, "To be named as co-chairman for Wildlife Week, along with Walt Disney, is a nice compliment, and I only wish that I could contribute something beside[s] my name to such a worthy cause."[45] Disney, selected for the honor because of his series of spectacular wildlife and nature movies, served alone. Darling died a month before National Wildlife Week began.

# 21

## " 'Bye Now—It's Been Wonderful"

THE LAST SEVERAL YEARS of Jay Darling's life were a cloud cover of sickness broken intermittently by bright periods of robust activity. In 1959 he was hospitalized for what he described as "ten poisonous days," and only later did he feel "as if I might be going to live."[1] It was apparently during that seige that Darling created his farewell cartoon. Darling's sight and hearing were failing, and he found it difficult to deal with the erosion of his senses. Darling recollected receiving a call from a friend while staying in northern Minnesota:

When I was summoned from my cottage on the edge of Long Lake to come to the Lodge to take a long distance call from Baltimore, it was such a surprise that I must have been temporarily knocked off my pins—so much so that I even forgot that I never answer telephone calls any more unless it's over one of the hard-of-hearing instruments, which no-one ever heard of in the northern lake district of Minnesota. I was so surprised that I forgot all about a sclerotic heart and shortness of breath, and hurried down to the Lodge, picked up the receiver just as I used to twenty-five years ago, and not until I found that I couldn't hear a damned thing did I realize that I wasn't living in that world any more, and that I must ask the Lodge-keeper to act as stand-in for me for the ensuing conversation.[2]

Darling often remarked on his fading memory, eyesight, and hearing. "Most of the glue," he lamented in February 1960, "has come off the old gummed labels in my memory." Darling's poor eyesight resulted from incipient cataracts, "which have been clouding my eyesight a little more each year for the last ten years." Ding conceded they were bothersome, but he was resigned to the probability that they would "last as long as I do." He was resigned as well to living the rest of his days without ever again visiting Washington, D.C. "My chief trouble is a shortness of breath and a weak heart," he wrote his old friend Clark Salyer. "I do all right and feel fine if I just put on my old carpet slippers and sit in a well padded rocking chair."[3]

Darling's pragmatism applied to his own condition, and it was expressed in a comment on a new book by William Vogt:

I have just read Bill Vogt's new book entitled "People" which is the best treatise on overpopulation that I have yet seen. He not only thinks people should voluntarily submerge their enthusiasm for more babies but suggests it would be all right with the world if oldtimers who have reached the period where they can contribute nothing to the welfare of the human race should commit suicide. Quite a doctrine! And I am quite reconciled to the principle if it were not for the stigma which it leaves in the minds of unrealistic sapheads. I certainly have been hanging on long after my time, and for no good purpose, and would welcome it if the Great Referee would blow the whistle on me.[4]

In the spring of 1961 Darling was disturbed by another serious illness, but this time it was afflicting his Penny. She had picked up an intestinal "bug" or arsenic poisoning from insect repellant used in New Zealand when the couple had toured there earlier that year. Darling noted, with a hint of submerged fury, that as early as 1936 the Biological Survey had issued a pamphlet spelling out the destructive consequences of using insecticides. He asked experts in Washington, D.C., to send related information to the physicians treating Penny at the Mayo Clinic. He accused the doctors of preferring to cure people after they were poisoned "rather than participating in its prevention." Penny finally was released but suffered aftereffects of the excessive administration of antibiotics used to rid her of what was presumed to be a tropical bug. Later, however, Penny was apparently on the road to recovery, and Darling conceded that both he and his wife were quite happy over the diagnosis and evident cure dictated by the Mayo Clinic.[5]

Later that year Darling was the victim of a stroke. The effects were devastating to the zealously independent, free-wheeling Ding:

I have been hit again by a misplaced arteriosclerosis, which left me with a floppy left wing and a fuzzy head. If it had put the finishing touch to me and freed me from contemplation of this world and its maladjustments I think it would have been better employed than just dimming my vision and paralyzing my hand. The Talking Books are my salvation. My hearing is so bad that I can't make out the voices of people who try to read to me and when you cut out the daily newspapers it's pretty dull business for an old newspaper man.[6]

Merle Strasser picked up the mail at the *Register* office and took it to Darling's apartment on West Grand Avenue where she read his correspondence to him. He dictated responses, which she took back to the office, transcribed, and put in the mail. Darling's chauffer and handyman, Matthew Johnson, sometimes brought Ding's mail to him also; and members of Darling's family read to him continually from newspapers, books, and magazines.[7]

Darling, plunged into darkness, silence, and confusion, was plunged into despondence as well. He wrote, "With me now, arteriosclerosis has destroyed my vision and made my left hand useless. That's the doctor's diagnosis. My mind is so fuzzy that I can't even tell what time it is when my watch is in front of my eyes. If there were only some way I could sign off without causing so many people a lot of trouble and unhappiness I'd sure take a quick way out." Merle Strasser felt the need to add a "secretary's note" to the transcription, explaining the tone of her boss's letter: "I read your letter to him, as I do all of the mail, and items from the daily papers, but I know how frustrating it is to him not to be able to read. He has always been quite uncomplaining of his lot but now that this last pleasure—reading—has been denied him he gets pretty low. Sorry."[8]

In a letter to Max McGraw, who had inquired about his health, Darling made no mention of the stroke. Merle again took it upon herself to add a note, this time to inform McGraw of Darling's condition and of Ding's deep appreciation for letters from old friends.[9]

She responded to another letter without consultation with Ding. She described Darling's stroke and its results:

Now, in addition to his greatly deteriorating vision, his equilibrium is wrecked and left hand numb, although his voice is as strong and clear as ever and he is not bedfast—can get around in their small apartment, but not outside. You can well imagine that one who has been active all his life would be quite frustrated at being sidelined and dependent on others. He wishes he could "walk out" quickly and easily and thus avoid becoming a burden to his family, a thing that worries him greatly.[10]

In January 1962 Ding wrote the widow of his lifelong friend, Paul Howe: "Since suffering a mild stroke last fall I have meditated so often on the fact that Mother Nature saves up a lot of physical frailties for oldtimers. How much better it would be if we could leave them all behind, along with Measles and Whooping Cough."[11]

For years prior to his debilitating stroke, Darling had spent many weekend hours at his Peony Farm west of Des Moines, where he kept the materials and equipment for his etching and worked at his favorite pastime. The farm was Darling's retreat. He could escape constant attention and involve himself in pure expression there, free of other disruptions or entanglements. It was his Walden—a very special place where he could be himself, surrounded by natural beauty; where he could enjoy solitude; or where he could enjoy the company of kindred souls. He would often have food catered in. "Ding liked to eat and we always had bottles out there," Gordon Meaney recalled. "Ding would say, 'You can't fly on one wing.'"

Jay and Penny had also spent many Sundays at the home of Forest and Fae Huttenlocher, looking after other artistic interests. Ding and Huttenlocher were members of the Board of Trustees of the Edmundson Memorial Foundation, later the Edmundson Foundation, Inc., which was charged with investing a substantial gift in the creation of an art center for Des Moines.[12] Members of the board, all prominent businessmen, were so busy in other interests that they set Sunday noon as their regular meeting time, and Darling persuaded Mrs. Huttenlocher to provide luncheons for a series of meetings.

At Fae's suggestion, several members of the Board visited with architect Eliel Saarinen, when other proposals for an art center design were judged too opulent for Iowa or too expensive for the board. The famous Saarinen was finally commissioned to design the Art Center. Although Darling was a prime mover in the drive for an art center in Des Moines, and even though he was responsible for the contribution of the magnificent Winnie Ewing Coffin art collection to the center, he eventually grew disgusted and resigned from the board. He apparently was miffed when his friend, James Earle Fraser, who had suggested a sculpture of Pegasus for the Art Center fountain was not commissioned to do the work. The Pegasus was done by Carl Milles instead. "Not much was said about it at the time," Fae Huttenlocher recalled, "but that was probably because the Pegasus that was done is generally regarded as outstanding." Darling also took exception to the acquisition policies of early Art Center directors, who preferred to concentrate on contemporary American art rather than on ancient works.[13]

Darling, to put it charitably, was not a fan of modern art. He considered it disgusting and decadent. When, during the Depression, young American artists were commissioned to paint murals in public buildings, Darling was incensed: "The worst part of the matter is that the type of mind which conceives such atrocities and the people who accept them and think they are art, seem to be in the vast majority in the United States and have taken over the government. . . ." Darling had little regard for art schools either: "Art schools are for mental cripples who can't stand on their own legs so have to be pushed around in a wheel-chair. Perhaps I am prejudiced; I never went to an Art School in my life—a statement on which Rollin Kirby once commented 'Your work certainly shows it but I wouldn't brag about it.' ''[14]

If he was hypercritical of the artwork of others, however, he was equally critical of his own cartoons and etchings. Darling's etchings were never sold by him. He rendered them as a hobby and a pastime, and his study of etching had convinced him that his works were not those of a master etcher. He distributed them freely to friends. "Hell!" he wrote an acquaintance, "If I had known that you wanted any of my duck etchings you would have had them long ago."[15] To another conservation colleague, he wrote:

The highest compliment anyone could pay me would be to want to possess some copies of my etchings. That is all I make them for and you need have no hesitancy about commandeering any copies of past etchings or those I may make in the future. I do not kid myself a bit about being a competent etcher. They lack much in professional technique and I can't let you go on cluttering up your walls with etchings without warning you that one of these days some high-grade art critic will come by and give you a humiliating rating as a judge of etchings.[16]

Darling's hideaway and his etching were realizations of a dream to which he had clung for many years. As early as 1920 he wrote a friend in Sioux City, "Painting outdoor scenes and particularly hunting pictures is one of the things I am always going to do but never get around to."[17]

Darling felt no more proprietary interest in his cartoons than in his etchings. Ralph Ellsworth, from the University of Iowa Library, had hounded Darling for years to deposit his large, original cartoons, from a lifetime of work, in the university library. Darling put him off, arguing that nobody would be interested in seeing the library devote substantial space to a collection of Darling's "cold potatoes."[18] Leslie Dunlap, the university's director of libraries, even offered to pay for the original cartoons. Darling responded:

Oh for heaven's sake, don't offer to pay for any of those old original cartoons! They aren't worth a dime a dozen and the attics of the U.S.A. are full of the works of "has-beens" presumed to have a post-mortem value but which, by the next generation, had drifted into the category of forgotten men. . . . I used to argue, hours at a time, with Ralph Ellsworth, about the folly of using valuable space in the University Libraries for the storage of such short-lived material as my cartoons and letter files.[19]

Darling freely gave his original cartoons to virtually anyone who made a specific request, on a first-come, first-served basis. He finally relented, however, and even helped Ellsworth remove the cartoons from the Register and Tribune Building and put them in the truck the librarian had driven from Iowa City. Even after the cartoons were deposited in the University of Iowa Library, however, Darling accommodated those who requested originals. As it was, the entire collection was reduced by more than half because of Ding's generous policy; yet Darling continued it. He suggested that inquirers go to Iowa City and tell the archivists they had Ding's permission to remove the requested cartoons. Dunlap finally wrote Darling and tactfully told him the practice would have to stop.[20]

Darling and the University of Iowa Library had collaborated earlier in a unique project that involved his cartooning skills and attracted national attention. Shortly after his retirement, Darling created a series of nine cartoon panels outlining the "history of education." The humorous series was roughed out by Darling as a joke, but Virgil Hancher, president of the university, and librarian Ellsworth were so amused that they urged Ding to execute the panels in finished form for etching in metal and mounting above the main doors to the university library. The project was the object of intense interest, but Darling wanted to be certain the *Register* had exclusive early access to the unique story. With the cooperation of John M. Henry and the library staff a subterfuge was arranged. The finished metal panels were temporarily mounted during a quiet weekend so the photographer from the *Register* could shoot his pictures. The panels were then removed until their permanent installation. As a result, the story first appeared in the Des Moines *Sunday Register Picture Magazine*. It appeared in *Look* magazine about two weeks later.[21]

Darling was a compulsive whittler. He dabbled in watercolors and oils and even became interested in sculpting. He took none of his means of artistic expression seriously, however. He wrote to a friend, "You can take your eye off the art exhibitions if what you are expecting to find is something from my workshop. I just putter around at odds and ends and have no serious objectives in mind. From now on my art work is largely a parallel occupation with the oldtimers who busy themselves with Solitaire."[22] One of Ding's cartoons remained to be exhibited, though, to the large audience served by the Des Moines *Register*.

Jay Darling suffered what was to be his final stroke early in February and died several days after he was hospitalized, Monday morning, February 12, 1962. An obituary appeared in the Des Moines *Tribune* that evening. The following morning, a lengthy article accompanied a simple, poignant cartoon titled " 'Bye Now—It's Been Wonderful Knowing You" on the front page of the Des Moines *Register*. Merle Strasser and Ken MacDonald had done their duty to Ding.[23]

# 22

## *Epilogue*

DOZENS of Jay Darling eulogies appeared in newspapers, magazines, and special-interest publications in the weeks and months following his death. The *Register* commented editorially on Ding's accomplishments and contributions, and *Register* columnist Harlan Miller wrote, "I happen to rank Jay Darling up there with Mark Twain & Will Rogers & Teddy Roosevelt & H. T. Webster & Pulitzer. He has the stature of the Great Ones, the Olympians, the Masters. This is a reminder how seldom we of this Great State fully recognize or acclaim our own!" Miller recollected Ding's "genial, stubborn, lovable, humane, dauntless, graceful charm" that touched those surrounding him with a special magic. He questioned why Ding had retired, recalled that Matisse painted into his eighties, and argued that Americans needed Ding as much as Matisse. "Maybe the answer is, Jay Darling never retired. He is a force transmissible from generation to generation. I see traces of him in the children of my daughter; she revered him. He'll still be with THEIR children in A.D. 2,000."[1]

Darling's transmitted force, his foresight, and his wisdom are pervasive spirits in a society that has begun to recognize the limits of its resources. On all sides there are reminders of his influence. There are also scars where some of Ding's warnings went unheeded.

Shortly after Darling's death, approximately sixty of his colleagues, admirers, and friends created the J. N. "Ding" Darling Foundation, Incorporated, as "a projection of Mr. Darling's great ideals." Officers and trustees of the Foundation included Sherry Fisher, Richard Koss, John Henry, Vernon Clark, Milton Caniff, Arthur Carhart, Robert Colflesh, Clarence Cottam, Gardner (Mike) Cowles, John Cowles, Charles K. Davis, Dwight D. Eisenhower, Ira N. Gabrielson, Seth Gordon, C. R. Gutermuth, Robert Herbst, Fletcher Knebel, Max McGraw, Fred and

Lewis B. Maytag, Fred Seaton, Carl Shoemaker, Harry S. Truman, James W. Wallace, Allen Whitfield, and dozens more. The Foundation wasted no time making its presence felt. By November 1962 its officers had enlisted the cooperation of federal agencies, Florida officials, and interested citizens in having the preserve on Sanibel Island designated the J. N. "Ding" Darling Wildlife and Waterfowl Sanctuary. In 1965 it was redesignated the J. N. "Ding" Darling National Wildlife Refuge, under the management of the Fish and Wildlife Service. The Darling Refuge was officially dedicated February 4, 1978, in special ceremonies on Sanibel Island.[2]

Sherry Fisher acted immediately to put the Foundation in the front rank for promotion of the Lewis and Clark Trail plan so enthusiastically proposed by Darling near the end of his life. Fisher and other representatives of the Foundation met with officials of the federal government, including Secretary of the Interior Stewart L. Udall, early in 1962. Udall and his subordinates were enthusiastic and supportive. They asked that the Foundation seek grass roots endorsement in the ten states traversed by the Lewis and Clark Trail. In October representatives of the states met in Portland, Oregon, at a session called jointly by the Darling Foundation and the Department of the Interior to consider the proposal. At a meeting the following month in Omaha, Nebraska, sixty-seven representatives of state and federal agencies, conservation organizations, historical societies, and others joined in several resolutions endorsing the plan to acquire additional land along the Missouri River, much of which was already publicly owned. The trail plan was not intended to create a park as such but anticipated the inclusion of historic sites, wildlife refuges, forest preserves, and public-use areas for hunting, fishing, boating, camping, hiking, and other outdoor activities.[3]

With such lively support from the affected states, the Department of the Interior took the proposal to Capitol Hill. The U.S. Congress in 1963 passed a concurrent resolution approving the trail plan. In October 1964 the Congress enacted Public Law 88-630, establishing the Lewis and Clark Trail Commission and providing that several members of the Darling Foundation be appointed members. Sherry Fisher served as commission chairman throughout its five-year lifetime, and the Darling Foundation continues to assist in the promotion and expansion of the Lewis and Clark Trail plan.[4]

Since its creation the Foundation has also provided annual grants to selected students in fisheries and wildlife management at Iowa State University. The grants are intended to underscore the need for communication skills on the part of conservationists. Darling, himself an articulate purveyor of scientific information wrapped in commonsense language, considered such skills crucial to public understanding and support of conservation programs.

The Foundation has assisted the Izaak Walton League in several of its endeavors. It has given financial assistance to the Natural Resources Council of America and to Ducks Unlimited in various national conservation and waterfowl preservation programs. It has provided additional support to the national refuge system. It has supported the Children's Forest in Des Moines, now named the J. N. "Ding" Darling Children's Forest. The Foundation supports a lecture series at Iowa State University, pertaining to environmental conservation, energy, and population problems. It has assisted tree-planting programs, water testing, and an inland water project. It has also aided water quality and conservation education conferences.[5]

The Foundation aided in the transfer of ownership of the largest single collection of Darling etchings to Iowa State University. The sixty-two original etchings, exquisite in their workmanship despite Darling's modest assessment of their worth, are on permanent display in the Scheman Continuing Education Building in the Iowa State Center, visited by approximately 60,000 persons attending courses and conferences there each year. Ding's etchings are mounted in the Darling Lounge, adjacent to the Brunnier Gallery, showcase for a variety of art exhibitions each year. The Foundation accumulated most of the etchings from Darling's friends, but the largest number by far came from Chloris Colflesh, widow of Darling's friend Robert Colflesh. Among the etchings displayed at the Scheman is Darling's design for the first federal duck stamp, accompanied by the first stamp, which was autographed by him.[6]

Darling's original cartoons also enjoy continuing popularity. The large drawings, housed at the University of Iowa Library, are occasionally loaned for exhibition in Iowa and elsewhere. A smaller collection of the original works, housed at the Department of History and Archives in Des Moines, are also shown from time to time. One of the most extensive exhibitions of Darling cartoons was presented in April and May 1977 in the Brunnier Gallery. The presentation, scheduled in connection with the university's student-sponsored Veishea celebration, featured more than forty of Darling's cartoons, selected and described by Darling's colleague John M. Henry. The Darling Foundation published a photographic album of the exhibition and distributed copies to several locations, including the Brunnier Gallery and the Department of Special Collections in the Iowa State University Library.[7]

John Henry in 1962 also completed a project he and Ding had

begun in 1961—a collection of cartoons representative of those executed by the cartoonist in his forty-nine-year career. When they began
the project, Darling suggested that the twosome devote three weeks to
it, working from about 10:00 A.M. until 2:00 P.M. each day. When they
had invested approximately three days and were up to about 1937 in the
chronological selection process, Darling glanced up at Henry, sighed
and said, "Oh hell, you can finish the book. I'm tired."

Selections for the book were made from engraver's proofs of Ding's
cartoons rather than from the full-size original drawings. (The proofs,
willed to John Henry by Ding, are housed at the Cowles Library at
Drake University in Des Moines. Many of the cartoons included in this
biography have been reproduced from the same engraver's proofs, with
the cooperation of Henry.) Darling made only two editorial changes
after Henry had completed his selection. Ding suggested that a wartime
cartoon depicting the German people as a beast be removed and that a
drawing of several youngsters following a baseball player be included
instead. In a foreword to the book, *Ding's Half Century,* Henry remarked on the continuing influence of Darling's cartoons:

> Ding's willingness to give away his originals has resulted in their ap-
> pearance at widespread and varied places. One is hanging around Buckingham
> Palace somewhere, and another—very significant in Ding's career—was on the
> wall of a fishing boat on the Mississippi River for years. Copies of another one
> are in the homes of all descendants of Teddy Roosevelt, and the original was at
> Oyster Bay for a long time. Herbert Hoover has many that Ding drew about
> him, and so did Al Smith. . . . A Catholic bishop asked for, and received a
> drawing on finances. And to Princeton University went one depicting modern
> war machines.[8]

Darling's influence is also recalled on Sanibel, which is waging the
battle for existence that Ding predicted it would. Darling and others
objected strenuously to the construction of a causeway from the
mainland to the island, but it was built anyway. The three-year struggle
to stop the bridge started a united organization, however, that went on
in the fall of 1974 to make Sanibel a free city, independent of Florida's
Lee County. The beaches where Anne Morrow Lindbergh was inspired
to write her *Gift from the Sea* are becoming covered by condominiums.
Thanks to the fact that Sanibel is the only Florida island with a
freshwater river and to the presence of the 4,306-acre Darling Wildlife
Refuge, birds still outnumber Sanibel's human residents by more than
one-hundred to one.

In 1973 the value of building permits issued for Sanibel in one
week exceeded that of those issued for all 1972. Today a new two-

bedroom beach house sells for $150,000, and the old line, "If God retired, He'd live on Sanibel" now prompts the sneering rejoinder, "Yeah, just so He could sell real estate on the side." In 1974 one of the fears expressed by Darling forty years earlier became a reality. Even though conservationists predicted serious consequences, construction of a trailer park was approved. Where Teddy Roosevelt speared manta rays, where Edna St. Vincent Millay wrote, and where Zane Grey fished, Gulf-front property brings more than $1,200 a front foot.[9]

Darling's transmitted force is evident in headlines that could have been written when Ding was waging two-fisted warfare on the despoilers of water, soil, and wildlife. In 1977, 250 million tons of fertile topsoil were being washed into the Gulf of Mexico every year by the Mississippi River. Half of all the topsoil in the Corn Belt states had been lost to water and wind erosion in just a century of farming. A million acres of prime agricultural land were being claimed each year by urban developers, dam builders, and miners. When the fifteen-year-old World Wildlife Fund held its first meeting in the United States in 1976, its focus was endangered species, including man himself. S. Dillon Ripley, chairman of the fund's U.S. appeal and secretary of the Smithsonian Institution, echoed Darling's concerns of a generation earlier: "For years conservation organizations were thought of as a sort of missionary area of the church, and conservation was merely akin to religious fervor. But fortunately those days are long gone. Conservation of nature today is inevitably involved with the problem of world stability and peace."[10]

The U.S. Soil Conservation Service (SCS), in the fall of 1977, called attention to the worst soil erosion since the depression years of the 1930s, due to overfarming and insufficient soil and water conservation measures. Associate administrator of the SCS, Norman A. Berg reported, "The great plow-up has put dust in the air, sediment in our water and piles of low-priced wheat in the streets."[11]

Two staff supervisors for the Iowa Conservation Commission reported that attempts by the Army Corps of Engineers to straighten and channelize the Missouri River along Iowa's western border had become an ecological disaster by late 1976. They revealed that Iowa had lost nearly 32,000 acres of its best wildlife and fish habitat along the river, with no prospect of ever getting it back. Where the river once meandered, creating oxbows and backwaters that fed underground water tables and provided prime resting areas for migratory waterfowl and spawning areas for many species of fish, the river flowed with such speed and power that it destroyed nearly all life in its channel. The scouring effect of the swiftly moving stream, they predicted, would put the river in a "ditch" twenty-six feet lower than the surrounding land within fifty years. Iowa's State Preserves Advisory Board in late 1977 made a plea for a comprehensive survey of remaining natural areas in Iowa, lest they all be destroyed in the following five years.[12]

On a broader scale, representatives of the National Audubon Society, the National Wildlife Federation, and the Wildlife Management Institute told a Senate committee that the government was doing a poor job in developing the wildlife and recreation potential of military bases and installations. The nation's military reserves, embracing approximately twenty-six million acres, included nearly as much land as the National Park System. Lynn A. Greenwalt, director of the Fish and Wildlife Service, disputed claims of insensitivity and reported on restocking and development projects on several military sites. The need "will always exceed the available management resources," Greenwalt lamented.[13]

The Iowa Conservation Commission, which Darling tried to insulate from political influence, in 1977 found itself facing charges of favoritism in the issuance of hunting privileges and in the use of commission-owned facilities. The commission was also accused of laxity in its enforcement of the law.[14]

Also in 1977 the National Park Service requested $100,000 for a study that might lead to the establishment of a national park in the loess hills of western Iowa. It was a move reminiscent of Darling's efforts to set aside land adjacent to the Missouri River and put it under federal protection. Darling would have applauded the fall 1977 final hearings on the fate of the Boundary Waters Canoe Area and indications that the unique million-acre wilderness in the northeastern tip of Minnesota would be accorded wilderness status. He would have approved of a series of statewide meetings to bring the story of "Land, Water and Energy for Iowa in Century III" to thousands of Iowans through Iowa State University's Extension Service and other organizations. Darling's original Cooperative Wildlife Research Unit at Iowa State University had grown by 1977 to seventeen wildlife units throughout the United States. The concept had been expanded to include fisheries in 1961, and by 1977 a total of forty-five fish and wildlife research units were in operation.[15]

One of Darling's greatest legacies descended from two pictures among the many thousands he drew. His first duck stamp illustration launched a program that since 1934 has raised more than $160 million for acquisition of approximately two million acres of prime waterfowl habitat. That land and the many millions of acres in other refuges in which he took a deep and lifelong interest are today marked by the familiar outline of the flying goose—the symbol of the National Wildlife Refuge system. Darling designed the symbol, and just as surely he designed the system as well.[16]

In the course of his long and active life Darling was repeatedly recognized for his stream of accomplishments. He received the Distinguished Service Award for Conservation from the trustees of Public Reservations in 1945. He was recognized for his contributions to the observance of Iowa's 1946 Centennial. He was awarded a life membership in the American Forestry Association and was given the Award of Merit by the National Physicians Committee for his support of the medical profession through his editorial cartoons. The Iowa State College Chapter of Sigma Delta Chi (now the Society of Professional Journalists, Sigma Delta Chi) elected Darling an honorary member in 1916, early in Ding's soaring career. Late in his life, Sigma Delta Chi designated the Register and Tribune Building as a historic site because of the distinguished cartooning career Ding had pursued there.[17] He also received the Des Moines Community Service Award and was honored by the Bison Society of America and the National Wildlife Association.

There were other awards—scores of them in addition to his Pulitzer prizes, honorary doctorates, the Audubon Medal, the Roosevelt Medal, and the prestigious Hutchinson Award. One is noteworthy because of his continued "participation" in it. The Iowa Award, created as an outgrowth of the state's centennial observance, is the state's highest honor. Its first recipient was Herbert Hoover in 1951, and its second was Jay Norwood Darling in 1956. Darling's sidekick John Henry helped arrange the surprise presentation to Darling at the national annual meeting of the Izaak Walton League in Sioux City. The 1961 award was presented jointly to Dr. James Van Allen and Dr. Frank Spedding for their leading roles in the study of space and atomic energy respectively. The 1965 award went to Henry A. Wallace, and the posthumous award was made in ceremonies on the Iowa State University campus. Mrs. Mamie Eisenhower received the award in 1970, and in 1975 it was made posthumously to Karl King, the famous Iowa composer of band music. The 1978 recipient of the Iowa Award was Norman E. Borlaug, Nobel Laureate, plant scientist, and leader of the green revolution credited by some with averting mass global starvation. The symbol given to Borlaug was an etching of canvasback ducks in flight—one of the most distinctive done by his fellow Iowa Award winner, Ding Darling. The etching, donated by the Darling Foundation, was presented to Borlaug at Iowa State University February 8, 1978.[18]

Darling's prominence as a political cartoonist has enjoyed vibrant durability. His artistic comments appear with regularity even in the most recent history texts. One of the most impressive signs of Darling's lasting influence can be found in a volume published in conjunction with the nation's bicentennial observance—*A Cartoon History of United States Foreign Policy, 1776-1976.* The book, compiled by the editors of the Foreign Policy Association, includes ten Darling cartoons. Only two cartoonists—Bill Mauldin and Pat Oliphant of contemporary vintage—are as well represented.[19]

John Henry, as much as any living person, has kept the influence and the memory of Ding the cartoonist before the eyes of an interested public for many years. Especially since Ding's death, Henry has been a tireless herald of the Darling legacy. Henry, better acquainted with the cartoonist's works than any other person, has faithfully responded for the Darling Foundation to the endless requests for information concerning Ding, his cartoons, and his etchings.

Darling's family, too, has kept the Ding spirit alive through active support of the Darling Foundation and involvement in ecological matters. Ding's daughter and her husband, Mary and Richard B. Koss of Des Moines, are Foundation stalwarts. Their son Christopher (Kip) Koss, a commercial pilot, is also a trustee of the Darling Foundation. (Dr. John Darling died December 19, 1973, at Lakeland, Florida. Penny Darling died December 13, 1968.)

Darling is also alive in the recollections of those who met and knew the man. He is remembered as a talented, energetic, humane person and as a mercurial, dynamic, and assertive personality. The admixture of attributes was attractive and exciting to most Darling acquaintances. As Gardner (Mike) Cowles, Jr., wrote, "He had a wonderful personality and had about as many friends and as few enemies as anyone I have known."

Darling's cartoons, profound in their simplicity, belied his complex, complicated makeup. A Teddy Roosevelt Republican, he became one of the most effective members of the Franklin Roosevelt New Deal team. An opponent of centralization in government, he saw the need for and used a big federal stick in his efforts to restore and protect land and wildlife. He devoted himself for years to the concept of a centralized organization of conservationists. Though he was a conservative in international politics, he nevertheless argued for U.S. support of the League of Nations and heartily endorsed the aims and purposes of the United Nations.

Darling's heyday coincided with that of the Hollywood star system. In a simpler time, marked by slower transportation and less sophisticated means of communication, national personalities were shrouded in mystique and placed on pedestals by an adoring public. Darling was never comfortable as the object of reverance and awe, but he could not escape his role as a public personality, who in a sense belonged to his country.

His interests were kaleidoscopic. They were seemingly endless in variety but were set in an ordered framework. Most were centered in conservation and grew out of Darling's compulsive quest for biological knowledge. With minor exceptions, Darling's multifarious involvements rested at some point on the circumference of the cycle of life.

As much as Darling derived satisfaction from the study of life, he relished the unbridled living of it as well. Darling's heritage was an uninhibited, wild frontier spirit. Like a mustang, he would not be hobbled; he would flee rather than be broken. Like the Indian pony he rode as a boy, he avoided civilization's trappings and its growing social complexities. His style of life was direct and uncomplicated. He was most at ease when he was in control. He was exceedingly capable in his many roles, and his strong personality led him to opt out of those groups that lurched in directions he chose not to take. "He was not," as Kenneth MacDonald put it, "very good as a team player." Whether in college, business, conservation, politics, or community service Ding took charge or he often took his leave.

The minister's son had a missionary's dedication. Deep in the gut of the hard-nosed crusader for self-reliance there lay an indigestible lump of idealism. The throbbing mass would not give him rest. It drove him toward utopia, but by a route different from that staked out by the impractical, intellectual idealists he so distrusted. Ding would make the world a more livable place and a more beautiful one. He would stop the waste of human life by curing illness and ending war. He would banish the mistreatment of animal and plant life. He would restore the ecological balance of Eden.

His standards approached perfection. He expected too much from himself and from other mortals. He was bound for disappointments and discouragements from the beginning. Yet despite his depreciation of his own efforts Ding's accomplishments were astounding. He may never have found utopia, but through his efforts subdivisions of Eden are preserved and protected for generations that might otherwise never have known that such natural beauty and balance ever existed.

# *Notes*

CHAPTER 1

1. New York *Globe,* Clipping, 1911; Scrapbook 2, Darling Papers (hereinafter DP).
2. The account of the early life of Marc Darling, before he entered the ministry, is from the privately published work, Marcellus Warner Darling, *Marcellus Warner Darling, 1844-1913* (Glencoe, Ill.: Clara W. Darling [1913]), pp. 7-23.
3. Marcellus Warner Darling, *Fusing Truth into Life* (Glencoe, Ill.: Marcellus W. Darling, 1906), pp. 12, 56.
4. M. W. Darling, *Darling,* p. 23.
5. Darling to Russell A. Runnells, Sept. 18, 1942, DP.
6. M. W. Darling, *Darling,* pp. 23-24.
7. *Saturday Evening Post,* Oct. 19, 1940, p. 34.
8. Darling to Lynn Bogue Hunt, Jan. 27, 1944, DP.
9. Darling to John A. Dehner, Jan. 26, 1959, DP.
10. Darling to Dr. and Mrs. Clarence Cottam, June 17, 1960, DP; Darling to Mrs. Addison Parker, Nov. 21, 1959, DP.
11. Darling to Dr. Stanley Young, Jan. 13, 1956, DP; Darling to Paul Errington, Jan. 4, 1958, DP; Darling to F. Fraser Darling, May 13, 1959, DP; Darling to J. Clark Salyer II, Dec. 7, 1960, DP.
12. Max McGraw, mimeo insert, 1960, DP.
13. Darling to Philip A. DuMont, Sept. 16, 1960, DP.
14. Darling to Hunt, Jan. 27, 1944, DP; *Post,* Sept. 21, 1935, pp. 14-15.
15. Darling to Errington, Jan. 4, 1958, DP.
16. Darling to M. O. Steen, July 28, 1960, DP; Darling to Young, Jan. 13, 1956, DP; Darling to Juanita Lines, Feb. 9, 1959, DP.
17. Handwritten notes, Scrapbook 1, DP.
18. Darling to Steen, Aug. 4, 1960, DP.
19. Darling to Lines, Feb. 9, 1959, DP.
20. Ibid.
21. Darling to Hunt, Jan. 27, 1944, DP; Darling to Charles H. Callison, July 1, 1947, DP.

CHAPTER 2

1. Darling to Lynn Bogue Hunt, Nov. 18, 1943, DP; *Current Biography,* July, 1942, p. 18.
2. D. R. Brown to Darling, May 5, 1921, DP; *Biography,* p. 18.
3. Darling to Hunt, Nov. 18, 1943, DP; *Saturday Evening Post,* Oct. 19, 1940, p. 34.

4. Darling to George W. Fenton, July 1, 1943, DP.
5. Steve Dougherty, Ding Darling, Fort Myers (Fla.) *News-Press,* Nov. 21, 1976, p. 5-D; Darling to Robert W. Howe, Nov. 17, 1941, DP.
6. Scrapbook 1, DP; Darling to Oscar P. Dix, May 26, 1948, DP; note, Scrapbook 1, DP.
7. Note, Theodore Lyman Wright to Darling, Mar. 21, 1898, Scrapbook 2, DP.
8. Dougherty, Ding; Walter Monfried, Ding Darling: His Pen Was His Sword, Milwaukee *Journal,* Feb. 25, 1971, p. 1; Sweet Retribution Comes to "Ding" after Quarter Century, Milwaukee *Journal,* July 12, 1925; When They Kicked "Ding" Out of College, Milwaukee *Journal,* June 27, 1928; B. J. Stager, An Interview with Jay Norwood Darling, *American Boy* [1917].
9. Darling to Charles Crutchfield, Dec. 19, 1941, DP; *Post,* p. 34.
10. *Biography,* p. 18; Dougherty, Ding.
11. Note, Scrapbook 1, DP; Darling to George X. Sand, Feb. 6, 1959, DP; Darling to Arthur Hawthorne Carhart, Sept. 19, 1960, DP; J. N. Darling: More Than a Cartoonist, *World of Comic Art,* June 1966, p. 19.
12. Darling to Vernon Carter, Apr. 10, 1945, DP.

## CHAPTER 3

1. Marcellus Warner Darling, *Marcellus Warner Darling, 1844-1913* (Glencoe, Ill.: Clara W. Darling [1913]), p. 26.
2. Darling to Mrs. Victor D. Vifquain, Aug. 9, 1943, DP; Walter Monfried, Ding Darling: His Pen Was His Sword, Milwaukee *Journal,* Feb. 25, 1971, p. 1; B. J. Stager, An Interview with Jay Norwood Darling, *American Boy* [1917].
3. *Who's Who in Iowa* (Des Moines: Iowa Press Assoc., 1940), p. 1273; *Annals of Iowa* 30(Jan. 1950):234; *The Story of Iowa: The Progress of an American State,* vol. 3 (New York: Lewis Hist. Publ., 1952), p. 40; John W. Carey, We Knew Him When, undated and unpaginated booklet, Scrapbook, DP; 50th Anniversary: Technical Journalism at Iowa State College, 1905-1955, Dept. Tech. J., 1955, p. 3.
4. A. F. Allen to Darling, Nov. 12, 1924, Allen Papers (hereinafter AP); Darling to R. H. Patchin, Oct. 22, 1948, DP.
5. Darling to George E. Bowers, Dec. 2, 1947, DP.
6. Scrapbook 1, DP.
7. Carey, We Knew Him, DP.
8. Monfried, Ding Darling, p. 1; Stager, Interview; Sioux City *Journal,* Mar. 28–May 31, 1901; Carey, We Knew Him, DP.
9. Darling to Elmer T. Peterson, May 18, 1953, DP.
10. Darling to Arthur Hawthorne Carhart, July 22, 1957, DP.
11. Darling to W. W. Waymack, Aug. 22, 1955, DP; Darling to Paul L. Errington, Jan. 4, 1958, DP.
12. *Saturday Evening Post,* Oct. 19, 1940, p. 34.
13. Scrapbook 1, DP.
14. W. H. Powell to Darling, Jan. 18, 1906, DP.
15. W. T. Buchanan to Darling, Sept. 6, 1906, DP.
16. John M. Henry to Lynette Pohlman, Apr. 4, 1977, property of Lynette Pohlman.
17. Darling to Vifquain, Aug. 9, 1943, DP; Scrapbook 1 and Family Scrapbook, DP; *Current Biography,* July 1942, p. 18.
18. Sioux City *Journal,* Apr. 4, 1901, p. 6.
19. Scrapbook 2, DP.

## CHAPTER 4

1. George Mills, *Harvey Ingham and Gardner Cowles, Sr.: Things Don't Just Happen* (Ames: Iowa State Univ. Press, 1977), pp. 3-5, 78.

2. Ibid.
3. Scrapbook 5, DP; Mills, *Ingham and Cowles,* p. 115; John Henry, A Treasury of Ding, *Palimpsest* 53(Mar. 1972):90–91.
4. Sioux City *Journal,* June 27, 1900, p. 1.
5. James B. Trefethen, *Crusade for Wildlife: Highlights in Conservation Progress* (Harrisburg, Pa.: Stackpole; New York: Boone and Crockett Club, 1961), pp. 1–3.
6. Henry, Treasury, p. 130.
7. Mills, *Ingham and Cowles,* p. 114; Henry C. Campbell to Darling, Mar. 15, 1910, DP; Charles H. Grasty to Darling, Apr. 11, 1910, DP; A. L. Clark to Darling, July 21, 1910, DP; C. R. Hope to Darling, Dec. 5, 1910, DP.
8. Marcellus Warner Darling, *Marcellus Warner Darling, 1844-1913,* (Glencoe, Ill.: Clara W. Darling [1913]), p. 26.
9. Darling to A. F. Allen, Jan. 25, 1913, AP; Scrapbook 2, DP; Des Moines *Register and Leader,* Nov. 2, 1911, p. 1.

## CHAPTER 5

1. Darling to H. T. Webster, Sept. 16, 1941, DP; Darling to W. W. Waymack, Aug. 31, 1955, DP; Scrapbook 2, DP.
2. Darling to George Matthew Adams, Nov. 18, 1943, DP.
3. Marcellus Warner Darling, *Marcellus Warner Darling, 1844-1913* (Glencoe, Ill.: Clara W. Darling [1913]), p. 28; Family Scrapbook, DP.
4. Edna Ferber to Darling, Nov. 12, 1912, DP.
5. George Mills, *Harvey Ingham and Gardner Cowles, Sr.: Things Don't Just Happen* (Ames: Iowa State Univ. Press, 1977), pp. 3–5, 78.
6. John M. Henry to Mrs. Cynthia B. G. Bush, Jan. 23, 1963, Special Collections, University of Iowa Library (hereinafter SCUI); Tintypes, Minneapolis *Tribune,* May 20, 1945; Scrapbook 5, DP; *Current Biography,* July 1942, p. 19; Darling to Graham Hunter, July 2, 1948, DP; Henry C. Campbell to Darling, Mar. 15, 1910, DP; Charles H. Grasty to Darling, Apr. 11, 1910, DP; A. L. Clarke to Darling, July 21, 1910, DP; C. R. Hope to Darling, Dec. 5, 1910, DP; Darling to Hunter, July 2, 1948, DP; B. J. Stager, An Interview with Jay Norwood Darling, *American Boy* [1917].
7. M. W. Darling, *Darling,* pp. 25–26, 28.
8. Mills, *Ingham and Cowles,* p. 116.
9. Darling to A. F. Allen, Jan. 25, 1913, DP.
10. Des Moines *Capital,* Oct. 28, 1911; Scrapbook 4, DP; W. B. Southwell to Darling, Oct. 22, 1912, DP; Lafayette Young to Darling, Sept. 16, 1912, DP.
11. Southwell to Darling, Oct. 22 and Dec. 5, 1912, DP.
12. Southwell to Darling, Jan. 7 and Jan. 11, 1913, DP.
13. Darling to Allen, Jan. 25, 1913, AP.
14. Gardner Cowles to Darling, Jan. 18, 1913, DP.

## CHAPTER 6

1. Scrapbook 1, DP.
2. Ibid.
3. *Current Biography,* July 1942, p. 14; George Mills, *Harvey Ingham and Gardner Cowles, Sr.: Things Don't Just Happen* (Ames: Iowa State Univ. Press, 1977), pp. 42–43.
4. Mills, *Ingham and Cowles,* p. 116; Joseph Pulitzer to Darling, May 10 and May 12, 1915, DP; 1917–1924 correspondence folder, DP; J. W. Carey to Darling, May 11, 1915, DP.
5. Scrapbook 1, DP.
6. Darling to Arthur B. Poinier, July 18, 1944, DP.
7. Darling to Mrs. Victor D. Vifquain, Aug. 9, 1943, DP.
8. Darling to Mrs. Peter W. Janss, Oct. 10, 1960, and May 19, 1961, DP.

9. John M. Henry, Ding Things, Apr. 4, 1977.
10. Family Scrapbook, DP.
11. Mills, *Ingham and Cowles,* p. 38.
12. John M. Henry to Mrs. Cynthia B. G. Bush, Jan. 23, 1963, correspondence files, SCUI.
13. Darling to A. F. Allen, Mar. 11, 1919, AP.
14. Ibid.

## CHAPTER 7

1. John M. Henry, A Treasury of Ding, *Palimpsest* 53(Mar. 1972):82–83.
2. Scrapbook 5, DP. A certificate of membership in the National Aeronautic Association of the U.S.A. is dated Mar. 17, 1925.
3. Scrapbook 1, DP; Des Moines *Evening Tribune-Capital,* July 20, 1927, p. 15; Anna G. Hubbell to Darling, Dec. 8, 1914, DP.
4. Seth Gordon, Ding Darling: Dynamic Conservation Leader, undated ms., p. 2, DP.
5. Darling to R. G. Townsend, Mar. 7, 1959, DP.
6. Darling to Townsend, Mar. 25, 1959, DP.
7. Steve Dougherty, Ding Darling, Fort Myers (Fla.) *News-Press,* Nov. 21, 1976, p. 5-D; Darling to A. F. Allen, Nov. 4, 1921, AP; Irma McGowan to A. F. Allen, Apr. 22, 1924, AP.
8. Henry, Treasury, pp. 85–87; Dougherty, Ding; John M. Henry, Ding Things, Apr. 4, 1977.
9. Darling to A. H. Carhart, Sept. 19, 1960, DP; A. H. Hume, The Playhouse of a Cartoonist, *House Beautiful,* Apr. 1924, pp. 482, 484; *Better Homes and Gardens,* Feb. 1926; Allen to Darling, June 3, 1929, AP; Scrapbook 2, DP.
10. Scrapbook 7, DP; Walter Monfried, Ding Darling: His Pen Was His Sword, Milwaukee *Journal,* Feb. 25, 1971, p. 1; Scrapbooks 1 and 7, DP.
11. Scrapbook 1, DP.
12. Henry, Ding Things, Apr. 4, 1977.
13. Ding, Greatest Serious Cartoonist, Had Little Talent, but He Toiled While Better Artists Loafed at Corner, Saint Paul *Pioneer Press,* July 13, 1924, Sect. 3.

## CHAPTER 8

1. Scrapbook 2, DP.
2. John M. Henry, Ding Things, Apr. 4, 1977.
3. Steve Dougherty, Ding Darling, Fort Myers (Fla.) *News-Press,* Nov. 21, 1976, p. 5-D; Clem F. Kimball and Walter H. Beam to Darling, Apr. 6, 1925, DP.
4. Sweet Retribution Comes to "Ding" after Quarter Century, Milwaukee *Journal,* July 12, 1925; *Current Biography,* July 1942, p. 14; Scrapbook 6, DP.
5. Saint Paul *Dispatch,* Apr. 20, 1925.
6. Darling to A. F. Allen, Nov. 9, 1925, AP; Allen to Darling, Nov. 10, 1925, AP; Darling to Allen, Nov. 12, 1925, DP.
7. New York *Herald Tribune,* Apr. 5, 1926; Scrapbook 2, DP.
8. Milwaukee *Journal,* Apr. 15, 1926.
9. Darling to Allen, Apr. 15, 1926, AP.
10. *Current Biography,* July 1942, p. 14; Scrapbook 2, DP; *Saturday Evening Post,* Oct. 19, 1940, p. 34.
11. Darling to Allen, Dec. 3, 1926, AP.
12. Log of a Winter Cruise and Scrapbook 1, DP.
13. Scrapbook 1, DP; Des Moines *Sunday Register,* Society News Sect., July 14, 1929, p. 1; Allen to Darling, Sept. 10, 1928, AP; Darling to Allen Sept. 19, 1928, AP.
14. Brief History of the MGC of Des Moines (from notes furnished by Mrs. Forest Huttenlocher), undated typescript, Men's Garden Clubs of America, Des Moines, Iowa.
15. Darling to Allen, July 12, 1926 and June 8, 1928, AP.

## CHAPTER 9

1. *Editor and Publisher,* July 12, 1930; Scrapbook 2, DP.
2. Scrapbook 1, DP.
3. Seth Gordon, Ding Darling: Dynamic Conservation Leader, undated ms., p. 2, DP; Arnold O. Haugen, History of the Iowa Cooperative Wildlife Research Unit, *Iowa Academy of Science* 73(1966):138.
4. John M. Henry, A Treasury of Ding, *Palimpsest* 53(Mar. 1972):83–84; Scrapbook 1, DP; John M. Henry, Ding Things, Apr. 4, 1977.
5. Darling to George Matthew Adams, July 29, 1944, DP.
6. Darling to Herbert Hoover, Jan. 30, 1930, DP.
7. Hoover to Darling, Feb. 8, 1930, DP.
8. Darling to Jack Miller, Aug. 4, 1961, DP; Darling to D. M. Pendleton, July 21, 1961, DP.
9. Clippings: J. N. Darling for Senator, Scrapbook, DP (most clippings included in this scrapbook are undated.); Sioux City *Journal,* May 30, 1931.
10. Clippings, DP; Marshalltown *Times-Republican,* June 1931; Sioux City *Journal,* June 1931.
11. Clippings, DP; Storm Lake *Pilot-Tribune,* Feb. 4, 1932.
12. Darling to Elmer T. Peterson, Sept. 21, 1956, DP.
13. Clippings, DP; Fred Davis, Convention in Chicago, Sioux City *Journal,* June 14, 1932; Scrapbook 5, DP.
14. Scrapbook 5, DP; Sioux City *Journal,* June 17, 1932.
15. Darling to Mrs. Rose M. Gregory, Oct. 28, 1942, DP.
16. Darling to Mrs. J. M. Moore, Oct. 28, 1942, DP.
17. Scrapbook 5, DP; Storm Lake *Pilot-Tribune,* June [18], 1932.
18. Clippings, DP.

## CHAPTER 10

1. Jay N. Darling, *Ding Goes to Russia* (New York: Whittlesey House, McGraw-Hill, 1932), p. vii.
2. Ibid., pp. 3, 50–52.
3. Family Scrapbook, DP.
4. Darling to Lucretia B. Smith Sherar, Aug. 26, 1954, DP.
5. Family Scrapbook, DP.
6. *Saturday Evening Post,* Oct. 19, 1940, p. 34.
7. Darling to Sherar, Aug. 26, 1954, DP; Death Notices, Des Moines *Register,* Sept. 24, 1952.
8. Darling to Sherar, Aug. 26, 1954, DP.
9. Roller Coaster, Des Moines *Register,* June 27, 1977, p. 2.
10. Darling to D. W. Turner, Dec. 8, 1961, DP.
11. Turner to Darling, Dec. 3, 1961, DP; Darling to Turner, Dec. 8, 1961, DP.
12. Arnold O. Haugen, History of the Iowa Cooperative Wildlife Research Unit, *Iowa Academy of Science* 73(1966):136–68; Seth Gordon, Ding Darling: Dynamic Conservation Leader, undated ms., pp. 2–3, DP; James B. Trefethen, *Crusade for Wildlife: Highlights in Conservation Progress* (Harrisburg, Pa.: Stackpole; New York: Boone and Crockett Club, 1961), p. 268.
13. Haugen, History, p. 139.
14. Ibid.; Gordon, Ding Darling, pp. 2–3, DP.
15. Darling to R. G. Townsend, Mar. 7, 1959, DP.
16. Gordon, Ding Darling, p. 4, DP.
17. Darling to Elmer T. Peterson, Sept. 21, 1956, DP.

## CHAPTER 11

1. Darling to A. D. Rathbone IV, July 20, 1946, DP; James B. Trefethen, *Crusade for Wildlife: Highlights in Conservation Progress* (Harrisburg, Pa.: Stackpole; New York: Boone and

Crockett Club, 1961), p. 264; Seth Gordon, Ding Darling: Dynamic Conservation Leader, undated ms., pp. 6, 8–9, DP.
2. Gordon, Ding Darling, pp. 8–9, DP.
3. Trefethen, *Crusade,* p. 263.
4. Gordon, Ding Darling, pp. 6, 8–9, DP.
5. Darling to Rathbone, July 20, 1946, DP; Darling to Clarence Cottam, June 25, 1959, DP.
6. Darling to Rathbone, July 20, 1946, DP.
7. Ibid.; Darling to Cottam, June 25, 1959, DP; Carl D. Shoemaker, The Stories behind the Organization of the National Wildlife Federation and Its Early Struggles for Survival, (Washington, D.C.: National Wildlife Federation, 1960), copy annotated and corrected by Darling, DP.
8. Darling to Cottam, June 25, 1959, DP.
9. Ding Flattered by Reported Kidnap Plot, Evansville (Ind.) *Journal,* Feb. 16, 1934; Des Moines *Register,* Feb. 16, 1934; Scrapbook 5, DP. The story was moved on the Associated Press wire service and appeared in many U.S. daily newspapers.
10. Gordon, Ding Darling, p. 8, DP.; Trefethen, *Crusade,* p. 264.
11. Darling to Clifford R. Hope, Aug. 28, 1937, DP.
12. Ibid.; Trefethen, *Crusade,* p. 264; Gordon, Ding Darling, p. 8, DP.
13. Shoemaker, Stories, p. 6; Darling to Rathbone, July 20, 1946, DP; Darling to Cottam, June 25, 1959, DP.
14. Gordon, Ding Darling, p. 11, DP.
15. Darling to Rathbone, July 20, 1946, DP.

CHAPTER 12

1. Darling to Clarence Cottam, June 25, 1959, DP.
2. Ibid.
3. Darling to Cottam, May 16, 1959, DP.
4. James B. Trefethen, *Crusade for Wildlife: Highlights in Conservation Progress* (Harrisburg, Pa.: Stackpole; New York: Boone and Crockett Club, 1961), p. 264; Wallace Names "Ding" U.S. Biological Chief, Milwaukee *Journal,* Mar. 10, 1934; J. N. Darling ("Ding") Named Chief of Biological Survey, Washington *Post,* Mar. 11, 1934 (the story was moved on the Associated Press wire service and appeared in many U.S. daily newspapers); U.S. Department of Agriculture, *Report of the Chief of the Bureau of Biological Survey* (Washington, D.C.: USGPO, 1934), pp. 1, 23; Darling to Elmer T. Peterson, Jan 25, 1954, DP; Trefethen, *Crusade,* p. 265.
5. Darling to Cottam, June 25, 1959, DP; Darling to A. D. Rathbone IV, July 20, 1946, DP.
6. Darling to Cottam, June 25, 1959, DP; Darling to Rathbone, July 20, 1946, DP.
7. Biographical Information, Clarence Cottam, May 28, 1948, copy of typescript with letter, Darling to Mrs. Roland Robinson, Aug. 16, 1961, DP.
8. Darling to Douglas McKay, May 22, 1953, DP; Darling to John Farley, Apr. 1, 1956, DP.
9. Promotions Announced by Interior's Wildlife Service, U.S. Dept. Int. news release, Sept. 12, 1961, DP.
10. Darling to Rathbone, July 20, 1946, DP.
11. Trefethen, *Crusade,* p. 272.
12. Darling to Rathbone, July 20, 1946, DP; Trefethen, *Crusade,* pp. 265–66.
13. Darling to Rathbone, July 20, 1946, DP.
14. Ibid.
15. Frank J. Rader, Harry L. Hopkins, the Ambitious Crusader, *Annals of Iowa* 44(Fall 1977):85, 90, 102.
16. Darling to Rathbone, July 20, 1946, DP; Seth Gordon, Ding Darling: Dynamic Conservation Leader, undated ms., p. 12, DP; Darling to G. Decker French, June 10, 1944, DP.
17. Trefethen, *Crusade,* p. 266.
18. Darling to Rathbone, July 20, 1946, DP.
19. Darling to Franklin D. Roosevelt, July 26, 1935, DP.
20. Roosevelt to Darling, July 29, 1935, DP.

## CHAPTER 13

1. Duck Stamps and Wildlife Refuges, U.S. Dept. Int., Fish and Wildlife Serv., Circ. 37, 1956.
2. Darling to J. Clark Salyer II, Sept. 2, 1954, DP.
3. James B. Trefethen, *Crusade for Wildlife: Highlights in Conservation Progress* (Harrisburg, Pa.: Stackpole; New York: Boone and Crockett Club, 1961), p. 267.
4. Darling to A. D. Rathbone IV, July 20, 1946, DP.
5. Quoted in Trefethen, *Crusade*, p. 268.
6. Ibid.
7. Ibid., p. 269.
8. Darling to M. Hartley Dodge, July 8, 1952, DP.
9. Darling to Charles K. Davis, Mar. 25, 1960, DP.
10. Darling to Dodge, July 8, 1952, DP; Darling to Nash Buckingham, Aug. 10, 1944, DP.
11. Trefethen, *Crusade*, p. 269; Arnold O. Haugen, History of the Iowa Cooperative Wildlife Research Unit, *Iowa Academy of Science* 73(1966):138.
12. Darling to Paul Cunningham, June 26, 1944, DP.
13. Darling to Clifford R. Hope, Aug. 28, 1937, DP.
14. Darling to James B. Trefethen, Apr. 5, 1960, DP.
15. Darling to Elmer T. Peterson, Jan. 18, 1954, DP.
16. Ibid.
17. Darling to Trefethen, Apr. 18, 1960, and Nov. 22, 1961, DP; Darling to Vernon L. Clark, May 25, 1960, DP.
18. Darling to Fred M. Packard, July 31, 1952, DP; Darling to Russell Lord, Feb. 23, 1960, DP.
19. Darling to Peterson, May 6, 1959, DP; Darling to Roy E. Hayman, Jan. 3, 1956, DP; Seth Gordon, Ding Darling: Dynamic Conservation Leader, undated ms., p. 14, DP.
20. Darling to Vernon Carter, Apr. 10, 1945, DP.
21. Gordon, Ding Darling, p. 15, DP; Trefethen, *Crusade,* p. 269.
22. Darling to Manly F. Miner, Aug. 23, 1955, DP; Darling to Salyer, Sept. 29, 1954, DP; Darling to Willis Robertson, Feb. 23, 1954, DP; Darling to W. R. Felton, July 23, 1959, DP.
23. Darling to Rathbone, July 20, 1946, DP.
24. Darling to Salyer, Nov. 8, 1935, DP.
25. Harold Ickes to Darling, Nov. 13, 1935, DP.
26. Scrapbook 5, DP; New York *Herald Tribune,* Nov. 12, 1935; Darling to Rathbone, July 20, 1946, DP.
27. Darling to Salyer, Sept. 30, 1953, DP.

## CHAPTER 14

1. Darling to Clarence Cottam, June 25, 1959, DP.
2. *Saturday Evening Post,* Oct. 19, 1940, p. 34; New York *Herald Tribune,* Sept. 6, 1935, p. 14, Sept. 9, 1935, p. 1.
3. "Ding," President's "Bad Boy," Returns to His Drawing Board, Milwaukee *Journal,* Nov. 18, 1935.
4. George Mills, *Harvey Ingham and Gardner Cowles, Sr.: Things Don't Just Happen* (Iowa State Univ. Press, 1977), pp. 117–18.
5. Ibid., pp. 117, 161.
6. Dr. Louis H. Valbracht, Service for Robert W. Colflesh, undated copy of typescript.
7. Elizabeth Cook, Robert W. Colflesh, campaign pamphlet [1934], p. 4.
8. Ibid., pp. 5–6.
9. Chloris Colflesh, Excerpts from Hawkeye Odd Fellow Magazine, undated typescript; Des Moines Loses an Outstanding Citizen, Des Moines *Tribune,* Apr. 18, 1967.
10. Darling Papers, SCUI.
11. Des Moines *Tribune,* Nov. 16, 1963, p. 14.
12. Mills, *Ingham and Cowles,* p. 117.
13. George P. Millington to Darling, Nov. 13, 1938, DP.
14. Darling to Irving Brandt, Jan. 6, 1936, DP.
15. Undated document, Item 3625, Folder 7-55, DP.

CHAPTER 15

1. James B. Trefethen, *Crusade for Wildlife: Highlights in Conservation Progress* (Harrisburg, Pa.: Stackpole; New York: Boone and Crockett Club, 1961), p. 269.
2. Seth Gordon, Ding Darling: Dynamic Conservation Leader, undated ms., p. 15, DP; Carl Shoemaker, The Stories behind the Organization of the National Wildlife Federation and Its Early Struggle for Survival, (Washington, D.C.: National Wildlife Federation, 1960), p. 1, copy annotated and corrected by Darling, DP; *Christian Science Monitor,* Feb. 7, 1936; *Current Biography,* July 1942, p. 14; Scrapbook 5, DP.
3. Darling to T. E. Doremus, H. P. Davis, and Seth Gordon, Dec. 9, 1935, DP; F. C. Walcott to Darling, Oct. 1, 1936, DP.
4. Lucy Furman to Darling, Jan. 9, 1938, DP; Darling to Furman, Jan. 11, 1938, DP.
5. Trefethen, *Crusade,* p. 270; Shoemaker, Stories, p. 7.
6. Trefethen, *Crusade,* p. 271.
7. Darling to Irving Brandt, Jan. 6, 1936, DP; Darling to E. A. Gilmore, Jan. 10, 1936, DP; The Iowa Lakeside Laboratory, State Univ. Iowa Bull. N.S. 902, Apr. 10, 1937, and N.S. 1031, Feb. 8, 1939.
8. Merle Houts Strasser to Kenneth Taylor, Apr. 16, 1936, DP; Arthur L. Clark to Darling, Feb. 23, 1937, DP; Strasser to Clark, Feb. 26, 1937, DP.
9. J. N. Darling, *The Cruise of the Bouncing Betsy: A Trailer Travelogue* (New York: Stokes, 1937), pp. 9, 15–16.
10. Ibid., p. 94; Mark Twombly, J. N. Ding, Sanibel and Captiva (Fla.) *Island Reporter,* Apr. 29, 1977, p. 1-B.
11. Darling, *Cruise,* pp. 93–94.
12. George Stevens to Darling, May 26, 1943, DP.
13. Telegram, Darling to E. A. Gilmore, Dec. 7, 1937, DP; Darling to Guy M. Gillette and Clyde Herring, Dec. 6, 1937, DP; Darling to Conrad Wirth, Dec. 4, 1937, DP; Telegram, Darling to Alf Landon, Aug. 27, 1937, DP; Darling to Clifford R. Hope, Aug. 28, 1937, DP; Darling to William Allen White, Aug. 28, 1937, DP; Landon to Darling, Sept. 2, 1937, DP.
14. Gordon MacQuarrie, Ding Darling—The Spirit of Conservation, Milwaukee *Journal,* Mar. 7, 1937.
15. Gordon MacQuarrie, Darling Steals Wild Life Show, Milwaukee *Journal,* Mar. 4, 1937; Darling Papers, 1935–1936.
16. MacQuarrie, Darling, Milwaukee *Journal,* Mar. 4, 1937.
17. Shoemaker, Stories, p. 20.
18. *Current Biography,* July, 1942, p. 14; Trefethen, *Crusade,* p. 271.
19. Darling's Son Injured, Auto Hits Bridge, Des Moines *Register,* Jan. 14, 1939, p. 1.
20. Merle Houts Strasser to Miles D. Kinkead, Jan. 23, 1939, DP.
21. Darling to Paul Cunningham, June 26, 1944, DP; Wildlife Conference, *Time,* Feb. 27, 1939, p. 41; Merle Houts Strasser to Harry J. Krusz, Mar. 8, 1939, DP.
22. Shoemaker, Stories, p. 21.

CHAPTER 16

1. Des Moines *Register,* June 14, 1940; Scrapbook 5, DP.
2. Darling to Mrs. [Mina Edison] Hughes, Feb. 12, 1941, DP; Darling to Hugh H. Bennett, Dec. 24, 1940, DP; Darling to Ira N. Gabrielson, Feb. 2, 1940, DP; Edison to Darling, Nov. 4, 1940, DP; Merle Houts Strasser to Edison-Hughes, Oct. 18, 1940, DP; Darling to Dr. Arthur E. Bestor, Mar. 11, 1940, DP.
3. Darling to Walter P. Taylor, Nov. 28, 1941, DP; Darling to Ross O. Stevens, Aug. 19, 1941, DP; Darling to Russell Lord, July 7, 1941, DP; Darling to Gardner Cowles, Sr., Jan. 14, 1941, DP.
4. Darling to Paul C. Howe, Nov. 5, 1942, DP.
5. Darling to Bertram Gumbert, Nov. 5, 1942, DP; Darling to T. F. Mitchell, Aug. 11, 1942, DP.
6. Darling to Elmo Roper, Aug. 18, 1943, DP; Darling to King Park, Sept. 23, 1942, DP.

7. Darling to Charles L. Horn, Oct. 28, 1942, DP.
8. Darling to W. W. Waymack, Feb. 25, 1944, DP.
9. George Mills, *Harvey Ingham and Gardner Cowles, Sr.: Things Don't Just Happen* (Ames: Iowa State Univ. Press, 1977), pp. 106-8, 158-59; John Cowles to Darling, Dec. 16, 1941, DP; Darling to J. Cowles, Oct. 21, 1942, DP; Darling to Gardner (Mike) Cowles, Jr., Aug. 13, 1942, DP; Darling to G. Cowles, Sr., Jan. 14, 1941, DP; Darling to Century Club, New York, Sept. 3, 1941 (Darling recommended Mike Cowles for membership at the suggestion of Wendell Willkie), DP; Darling to Grantland Rice, May 6, 1943, DP.
10. Darling to Mrs. Samuel I. Rosenman, May 11, 1943, DP; Mills, *Ingham and Cowles*, p. 159; Darling to Herbert S. Houston, Nov. 10, 1943, DP; G. Cowles, Sr., to Darling, Apr. 5, 1943, DP.
11. G. Cowles, Sr. to Darling, Dec. 31, 1940, DP; Darling to G. Cowles, Sr., Dec. 5, 1941, DP.
12. Darling to William Hard, May 17, 1943, DP; Robert L. Ripley to Darling, May 6, 1943, DP.
13. Darling to Charles T. White, May 7, 1943, DP.
14. Gifford Pinchot to Darling, Sept. 10, 1943, DP; Darling to Pinchot, Sept. 17, 1943, DP.
15. Darling to E. E. Lincoln, Sept. 17, 1941, DP; Darling to R. E. Overholser, July 14, 1942, DP; Darling to George W. Fenton, July 8, 1942, DP.
16. Darling to Guy M. Gillette, Sept. 8, 1943, DP; Darling to Fred J. Poyneer, Sept. 21, 1943, DP.
17. Darling to Clarence E. Hall, Aug. 26, 1942, DP; Mark Twombly, J. N. Ding, Sanibel and Captiva (Fla.) *Island Reporter*, Apr. 29, 1977, pp. B-14, 15; Darling to Mrs. Carl Miner, June 25, 1942, DP; Darling to Spessard Holland, Dec. 22, 1943, DP; Holland to Darling, Jan. 19, 1944, DP.
18. Mrs. Virginia M. Mannon to Darling, Aug. 18, 1942, DP; Darling to Mrs. Harry Stanwood, Aug. 2, 1941, DP.
19. Darling to William C. Moore, May 31, 1940, DP.
20. Darling to Charles W. Collier, June 22, 1940, DP.
21. Darling to Lord, Mar. 18, 1942, DP; Darling to Edward K. Love, Feb. 18, 1942, DP; Darling to Gabrielson, Mar. 17, 1942, DP.
22. Darling to Harold L. Ickes, June 3, 1943, DP.
23. Darling to James G. Mitchell, Aug. 21, 1942, DP.
24. Darling to David A. Aylward, July 31, 1941 and Nov. 17, 1942, DP; copy of Aylward to Chiles P. Plummer, Nov. 12, 1941, DP.
25. Folder 4-22 and F. 4-22, Item 2251, DP; Steve Weinberg, Hubbell Estate Valued at $100 Million, Des Moines *Sunday Register,* Oct. 30, 1977, p. F-1.
26. Darling to Sam B. Rosenbaum, Nov. 5, 1942, DP; Darling to Rosenbaum, Oct. 20, 1942, DP; Folder 4-23, DP.
27. Darling to Sam Stevens, Aug. 1, 1942, DP; Darling to Mary Nye Hayes, Aug. 22, 1949, DP; Darling to Frank Knox, July 24, 1942, DP; Darling to J. Cowles, Dec. 9, 1942, DP; Darling to Percy Haven, July 7, 1943, DP; Darling to "Dr. Buie," Nov. 5, 1942, DP; Darling to Clarence Cottam, Aug. 14, 1945, DP.
28. Darling to Paul J. Nowland, June 21, 1943, DP.
29. Darling to William Vogt, Aug. 17, 1945, DP.
30. Darling to Fae Huttenlocher, Oct. 13, 1942, DP.

CHAPTER 17

1. Darling to David A. Aylward, Feb. 11, 1946, DP.
2. W. B. Cartwright to Darling, July 8, 1946, DP; Darling to Cartwright, July 11, 1946, DP.
3. Darling to Aylward and Miss Dorothy Gleeson, Mar. 19, 1947, DP.
4. Darling to Aylward, Mar. 5, Mar. 8, and Apr. 17, 1946, DP.
5. Darling to Chiles P. Plummer, Apr. 15, 1946, DP; Darling to Aylward, Apr. 17, 1946, DP.
6. Darling to Kenneth D. Morrison, Sept. 17, 1945, DP; Darling to Nash Buckingham, Oct. 5, 1944, DP.
7. Darling to Henry M. Magie, July 26, 1945, DP; Darling to Morrison, Sept. 27, 1945, DP.
8. Darling to William Vogt, Sept. 18, 1945, DP; Darling to Buckingham, Jan. 30, 1946, DP.

9. Darling to Daniel B. Beard, June 10, 1946, DP.
10. Darling to Dr. H. H. Brierly, Aug. 23, 1946, DP.
11. Darling to Harry Boyd, Aug. 29, 1946, DP.
12. Darling to Robert Blue, July 20, 1946, DP; Blue to Darling, Aug. 22, 1946, DP.
13. Darling to Blue, Aug. 22, 1946, DP.
14. Darling to F. W. Beckman, Sept. 12, 1946, DP; Darling to Dwight Bannister, Sept. 4, 1946, DP; Darling to Don L. Berry, Sept. 11, 1946, DP.
15. Darling to Blue, Nov. 21, 1946, DP.
16. Darling to Ira N. Gabrielson, May 25, 1946, DP.
17. Darling to Harold L. Ickes, Jan. 26, 1946, DP; Gabrielson to Darling, Jan. 29 and Feb. 14, 1946, DP; Darling to Augustus S. Houston, Mar. 25, 1946, DP; Darling to Clarence Cottam, Mar. 26, 1946, DP.
18. Darling to Guy Hinkley, Sept. 5, 1946, DP.
19. Darling to Miss Harlean James, Feb. 26, 1946, DP.
20. Darling to Bourke B. Hickenlooper, Mar. 18, 1946, DP.
21. Darling to A. F. Allen, Mar. 8, 1946, AP; Darling to Charles H. Callison, July 1, 1947, DP.
22. Darling to Bernard Baruch, May 25, 1948, DP.
23. Darling to James S. Russell, Mar. 26, 1946, DP.
24. Darling to Verne Marshall, Mar. 26, 1946, DP; Darling to W. W. Waymack, Jan. 12, 1946, DP.
25. Darling to Fred J. Poyneer, Dec. 5, 1947, DP; Darling to Frank W. Mattes, July 9, 1947, DP.
26. Darling to Art W. Smith, Nov. 2, 1948, DP.
27. Resolutions for IWLA, Sept. 24, 1948, DP.
28. Darling to Harvey Ingham, Sept. 10, 1948, DP.
29. Darling to Frank Thone, Sept. 27, 1948, DP.
30. Darling to George W. Dulany, Jr., Oct. 4, 1948, DP.
31. Darling to James Earle Fraser, Aug. 13, 1948, DP; Helen Rogers Reid to Darling, Nov. 5, 1948, DP.
32. Darling to Dulany, Oct. 4, 1948, DP; Des Moines *Sunday Register,* Apr. 24, 1949, p. 1; *Newsweek,* Apr. 25, 1949, p. 60.

## CHAPTER 18

1. Darling to J. A. Krug, July 6, 1949, DP.
2. Darling to Mrs. Frank Hayes, Aug. 22, 1949, DP.
3. Ibid.
4. Ibid.
5. Darling to Chiles P. Plummer, Aug. 18, 1949, DP; James B. Trefethen, *Crusade for Wildlife: Highlights in Conservation Progress* (Harrisburg, Pa.: Stackpole; New York: Boone and Crockett Club, 1961), pp. 230, 332; Darling to Archibald B. Roosevelt, Aug. 20, 1949, DP.
6. Darling to Gerald F. Baker, Aug. 22, 1949, DP.
7. Darling to William Vogt, July 22, 1948, DP.
8. Jay N. Darling, Conservation Education, typescript, Sept. 15, 1950, DP.
9. Darling to Roy E. Simpson, Sept. 19, 1950, DP; Darling to Gordon Smith, June 30, 1952, DP.
10. Darling to Peter Kuyper, June 20, 1942, DP; Darling to Charles F. Martin, Feb. 4, 1949, DP.
11. Darling to Mrs. Clarence Avery, Nov. 28, 1941, DP; Darling to William F. Russell, June 9, 1942, DP; Darling to William L. Finley, Sept. 8, 1942, DP; Darling to David A. Aylward, Feb. 11, 1946, DP; Darling to Elmer T. Peterson, May 28, 1949, DP; Darling to Baker, June 11, 1949, DP; Darling to Worth McClure, Jan. 30, 1950, DP.
12. Darling to Aldo Leopold, Oct. 3, 1947, DP.
13. Darling to Vogt, July 22, 1948, DP; Darling to Paul D. Dalke, July 19, 1948, DP; Darling to Peterson, Oct. 5, 1951, DP.
14. Darling to Fred J. Poyneer, Oct. 1, 1946, DP.
15. Jay N. Darling, Second Version of Introduction, typescript draft with Darling to J. Clark Salyer II, Apr. 26, 1949, DP.

16. Darling to Oscar Chapman, Feb. 4, 1947, DP.
17. Jay N. Darling, Early Observations of the "Red Tide" or Fish Plague in Florida, undated Item 2702, DP; Darling to Paul S. Galtsoff, Feb. 17, 1947, DP; B. F. Ashe to Darling, Feb. 18, 1947, DP; Chapman to Darling, July 24, 1947, DP; Darling to Tom Wallace, July 28, 1947, DP.
18. Darling to Lloyd C. Stark, Nov. 12, 1952, DP.
19. Darling to Milton Eisenhower, Dec. 4, 1952, DP; Darling to Salyer, Dec. 1952, DP.
20. Darling to Baker, June 11, 1949, DP; Darling to Hillory A. Tolson, Aug. 17, 1949, DP.
21. Darling to Baker, Aug. 22, 1949, DP.
22. Darling to W. D. Wood, Oct. 15, 1949, DP; Darling to Herbert R. Mills, Oct. 23, 1949, DP; Darling to James Silver, Nov. 7, 1949, DP; Wood to Darling, Oct. 20, 1952, DP.
23. Darling to Charles E. Bennett, June 23, 1950, DP; Darling to A. B. Roosevelt, June 30 and Dec. 6, 1950, DP.
24. Darling to Peterson, July 16, 1949, DP.
25. Darling to Peterson, Oct. 15, 1949, DP; Darling to R. H. Musser, June 19, 1951, DP.
26. Darling to Basil L. Walters, Aug. 6, 1952, DP.
27. Jay N. Darling, In re: Red Rock Reservoir: Howell Site, typescript, Sept. 26, 1950, DP.
28. Darling to Clarence Cottam, July 24, 1951, DP; Darling to I. T. Bode, Oct. 12, 1951, DP; Darling to Plummer, June 19, 1951, DP.
29. Darling to Devereux Butcher, July 8, 1952, DP; Darling to Anthony W. Smith, Aug. 15, 1952, DP; Darling to Peterson, Aug. 9 and Oct. 14, 1952, and Oct. 20, 1953, DP.
30. Darling to Kenneth D. Morrison, Aug. 11, 1950, DP; Darling to Peterson, Oct. 4, 1952, DP.
31. Darling to Ivah Green, Aug. 11, 1950, DP; Darling to Bruce F. Stiles, Oct. 20, 1949, DP; Darling to Ewald G. Trost, Oct. 21, 1949, DP.
32. Darling to William Hard, Nov. 9, 1953, DP.

## CHAPTER 19

1. E. Mallinckrodt to Darling, Feb. 9, 1955, DP; Darling to Douglas McKay, Jan. 18, 1954, DP; Darling to Lloyd C. Stark, Oct. 27 and Nov. 12, 1952, DP; Darling to Elmer T. Peterson, June 2, 1954, DP.
2. Darling to Carl S. Miner, July 7, 1954, DP.
3. Darling to Allen T. Weeks, Aug. 10, 1954, DP; Darling to Mallinckrodt, Nov. 9, 1954, DP; Darling to Thomas E. Martin, Henry Talle, H. R. Gross, K. M. LeCompte, Paul Cunningham, James Dolliver, Ben F. Jensen, and Charles B. Hoeven, Dec. 16, 1954, DP.
4. Darling to Clarence Cottam, Feb. 10, 1954, DP.
5. Darling to Herbert Hoover, Mar. 22, 1954, DP.
6. Darling to William J. O'Brien, Sept. 2, 1954, DP; Darling to Willis Robertson, Apr. 30, 1954, DP.
7. Darling to Bourke B. Hickenlooper, Apr. 30, 1954, DP.
8. Darling to McKay, July 12, 1954, DP.
9. Telegram, Darling to Frederick C. Lincoln for Cottam, July 14, 1954, DP.
10. Darling to Cottam, July 30, 1954, DP.
11. Darling to W. M. Apple, Mar. 24, 1954, DP; Des Moines Register, Sept. 27, 1955; Darling to C. R. Gutermuth, Sept. 14 and Sept. 28, 1955, DP.
12. Darling to Nash Buckingham, July 11, 1951, DP; Darling to James A. Ott, Aug. 20, 1951, DP; Darling to Devereux Butcher, Oct. 6, 1953, DP.
13. Darling to Charles S. Thomas, Sept. 21, 1955, DP.
14. Darling to John W. Tobin, Oct. 5, 1955, DP.
15. Darling to Charles K. Davis, Jan. 16, 1956, DP.
16. Darling to McKay, Jan. 20, 1956, DP; Darling to Mr. and Mrs. W. D. Wood, Jan. 28, 1956, DP.
17. Darling to J. Clark Salyer II, July 24, 1956, DP; Darling to Peterson, Sept. 21, 1956, DP.
18. Memo, Darling to Ries Tuttle, Oct. 1, 1956, DP.
19. Darling to R. W. Burwell, July 31, 1957, DP.
20. Darling to James R. Harlan, Feb. 27, 1958, DP; Darling to Conrad L. Wirth, July 9, 1958,

DP; Darling to Fred E. Hornaday, Aug. 19, 1957, DP; Darling to Cottam, Feb. 20, 1958, DP; Darling to Hubert H. Humphrey, July 22, 1957, DP.

21. Darling to Cottam, June 2, 1958, DP.
22. Darling to Miner, Sept. 11, 1958, DP.
23. Darling to Univ. Miami Marine Laboratory, Aug. 10, 1954, DP; Darling to J. Walton Smith, Oct. 15, 1954, DP; Darling to Miner, Aug. 17, 1954, DP.
24. Darling to Cottam, Aug. 24, 1955, and Feb. 27, 1958, DP; Darling to H. H. Shomon, Dec. 7, 1955, DP.
25. Darling to Lincoln, Nov. 13, 1956, DP.
26. Darling to Cottam, Apr. 8, 1958, DP.
27. Darling to Chiles P. Plummer, Aug. 4, 1955, DP.
28. Darling to Salyer, Oct. 12, 1955, DP.
29. Darling to Carl D. Shoemaker, Feb. 15, 1955, DP.
30. Darling to E. Laurence Palmer, Dec. 6, 1956, DP.
31. Darling to Richard H. Pough, July 19, 1956, DP.
32. Darling to Pough, Aug. 13, 1956, DP; Darling to Harlan, Feb. 22, 1957, DP.
33. Darling to Miner, July 7, 1954, DP; Darling to Ralph G. Cooksey, July 9, 1954, DP; Darling to Salyer, July 12, 1954, DP; Darling to Plummer, July 20, 1954, DP; Darling to Peterson, Jan. 18, 1954, DP; Darling to Cottam, Mar. 20, 1954, DP; Darling to Hugh B. Woodward, Aug. 16, 1954, DP.
34. Darling to Stanley P. Young, July 8, 1954, DP; Darling to Wirth, Nov. 30, 1954, DP; Darling to Harold Titus, Aug. 3, 1955, DP; Darling to Peterson, Sept. 17, 1956, DP.
35. Darling to Peterson, Oct. 28, 1955, DP; Darling to Cottam, Oct. 9, 1956, DP.
36. Darling to Wood, July 17, 1956, DP; Darling to Mr. and Mrs. B. F. Elbert, Jan. 5, 1957, DP; Darling to Gerald F. Baker, July 17, 1957, DP; Darling to Salyer, July 24, 1957, DP; Darling to C. E. Gillham, Nov. 21, 1957, DP.
37. Merle Houts Strasser to Fred E. Hornaday, Sept. 12, 1957, DP; Darling to Peterson, Aug. 27, 1958, DP; Darling to Thomas E. Martin, Dec. 8, 1958, DP.
38. Darling to Edward B. Wilber, Sept. 19, 1957, DP.
39. Darling to Wood, Sept. 10, 1951, DP.
40. Darling to Salyer, Aug. 3, 1955, DP; Darling to I. T. Bode, Oct. 3, 1955, DP.
41. Darling to Cottam, Apr. 27, 1957, and Nov. 26, 1958, DP; Darling to Bill Riaski, Sept. 11, 1958, DP; Darling to John A. Dehner, Dec. 9, 1958, DP.

CHAPTER 20

1. Darling to C. M. Frudden, Aug. 6, 1959, DP; Darling to J. Clark Salyer II, Oct. 3, 1955, DP.
2. Darling to Herschel C. Loveless, July 16, 1959, DP.
3. Darling to Louise L. Parker, July 16, 1959, DP; Parker to Darling, July 1959, DP.
4. Darling to Elmer T. Peterson, July 20, 1959, DP.
5. Darling to George M. Foster, Aug. 4, 1959, DP; Darling to Roy A. Stacey, Nov. 16, 1959, DP.
6. Darling to C. R. Gutermuth, Aug. 5, 1959, DP; Darling to James R. Harlan, Aug. 10, 1959, DP.
7. Darling to Stacey, Nov. 16, 1959, DP; Darling to J. D. Lowe, Sept. 21, 1959, DP.
8. Darling to Lowe, Sept. 21, 1959, DP; Darling to Foster, Sept. 16, 1960, DP.
9. Darling to Fred J. Poyneer, Aug. 28, 1959, DP; Darling to Sherry R. Fisher, Sept. 2, 1959, DP.
10. Darling to Bill Riaski, Sept. 29, 1959, DP; Darling to Foster, Aug. 10, 1959, DP.
11. Darling to Glen G. Powers, Sept. 2, 1960, DP; Darling to Norman A. Erbe, Nov. 16, 1960, DP.
12. Darling to Erbe, July 13, 1961, DP.
13. Erbe to Darling, July 14, 1961, DP; Darling to Erbe, July 18, 1961, DP; copy, Foster to Erbe, Jan. 26, 1961, DP; memo, Darling to Ries Tuttle, Sept. 12, 1961, DP.
14. Darling to A. C. Heyward, May 27, 1959, DP; Salyer to Darling, Oct. 23, 1959, DP; Darling to Clarence Cottam, Oct. 7, 1959, DP; Darling to Ross Leffler, Oct. 21, 1959, DP; Darling to Frank Gregg, Nov. 12, 1959, DP; Darling to John H. Baker, Jan. 8, 1960, DP.

15. Darling to Richard M. Nixon, May 12, 1960, DP.
16. Darling to Nixon, Mar. 29 and Apr. 18, 1960, DP.
17. Darling to Harlan, July 28, 1960, DP.
18. Darling to Cottam, July 28, 1960, DP.
19. Darling to Robert T. Howard, Sept. 1, 1960, DP; Darling to Nash Buckingham, Oct. 28, 1960, DP.
20. Telegram, Darling to John F. Kennedy, Henry M. Jackson, and Clark Clifford, Nov. 18, 1960, DP.
21. Darling to Stewart L. Udall, Dec. 7 and Dec. 28, 1960, DP.
22. Darling to Gutermuth, Dec. 13, 1960, DP; Cottam to Darling, Jan. 25 and Feb. 13, 1961, DP.
23. Darling to Clinton P. Anderson, June 6, 1961, DP.
24. Darling to Barry Goldwater, Quentin N. Burdick, Lee Metcalf, Frank Church, Henry Dworshak, and Ernest Gruening, June 6, 1961, DP; Darling to Anderson, July 13, 1961, DP; Telegram, Darling to Jack Miller, July 12, 1961, DP.
25. Darling to Bourke B. Hickenlooper, July 17, 1961, DP.
26. Anderson to Darling, July 14, 1961, DP; Darling to Anderson, July 18, 1961, DP; The Wilderness Bill, Des Moines *Sunday Register,* July 16, 1961, p. 14-G.
27. Darling to Peterson, June 22, 1959, DP.
28. Darling to Mrs. Dorothy R. Wade, June 26, 1959, DP.
29. Darling to Riaski, June 29, 1959, DP.
30. D. M. Area Dam Funds Voted, Des Moines *Register,* July 10, 1959, p. 1; Darling to M. O. Steen, July 28, 1960, DP.
31. Darling to Steen, July 18, 1960, DP.
32. Darling to Charles L. Horn, Aug. 18, 1960, DP.
33. Darling to Philip A. DuMont, Sept. 16, 1960, DP.
34. Darling to Fisher, Jan. 7, 1960, DP.
35. Darling to Robert McConnell Hatch, July 13 and July 27, 1961, DP; Darling to Ira N. Gabrielson, Aug. 9, 1961, DP.
36. John T. Eastlick to Darling, May 18, 1961, DP; Darling to Eastlick, June 5, 1961, DP; Darling to Floyd B. Chapman, Apr. 26, 1961, DP.
37. E. Laurence Palmer, Natural History Changes, *Nature,* Mar. 1959, pp. 137–44.
38. Darling to Charles K. Davis, Apr. 18, 1960, DP.
39. Darling to C. M. Goethe, Nov. 11, 1960, DP; Merle Houts Strasser to Frank Gregg, Feb. 8, 1961, DP; Darling to Paul Everett, Oct. 10, 1960, DP; Darling to Alfred O. Hoyt, Sept. 21, 1960, DP.
40. Erwin D. Sias, Journal Cartoon That Launched Ding's Career Became His "Last" by Quirk of History Writing, Sioux City *Sunday Journal,* Feb. 12, 1967, p. B-11.
41. Darling to Max McGraw, Dec. 7, 1960, DP.
42. Darling to Mrs. C. R. Gutermuth, Nov. 7, 1960, DP; Darling to Riaski, June 30, 1960, DP; Darling to Cottam, Nov. 10, 1960, DP; Darling to Gayer G. Dominick, Nov. 7, 1960, DP; Darling to Mrs. Gladys Crandall, Sept. 1, 1960, DP; Ding Honored at Dinner, Des Moines *Register,* Aug. 23, 1960, p. 3; Darling to Fairman R. Dick, Nov. 18, 1959, DP.
43. Darling to A. Willis Robertson, Oct. 5, 1960, DP; DuMont to Darling, Oct. 21, 1960, DP; Darling to Leffler, Sept. 19, 1960, DP; Darling to Horn, Sept. 22, 1960, DP.
44. Darling to Sidney L. James, Aug. 22, 1960, DP; Darling to Frederick C. Lincoln, Nov. 2, 1959, DP; Darling to Frank H. Mendell, Aug. 10, 1960, DP; Darling to Miss Juanita Lines, Feb. 9, 1959, DP.
45. Ries Tuttle, Darling and Disney Honored, Des Moines *Tribune,* Oct. 12, 1961, p. 21; Darling to Harlan, Oct. 19, 1961, DP.

CHAPTER 21

1. Darling to Mrs. Addison Parker, Sept. 21, 1959, DP.
2. Darling to Lee H. Hoffman, Sept. 29, 1959, DP.
3. Darling to C. R. Gutermuth, Feb. 2, 1960, DP; Darling to Elmer T. Peterson, July 5, 1960, DP; Darling to J. Clark Salyer II, Sept. 15, 1960, DP.

4. Darling to Carl D. Shoemaker, Oct. 26, 1960, DP.
5. Darling to Arthur H. Carhart, Apr. 20, and May 12, 1961, DP; Darling to James R. Harlan, Oct. 19, 1961, DP; Darling to Dr. and Mrs. Clarence Cottam, May 23, 1961, DP; Darling to Cottam, June 7, 1961, DP.
6. Darling to Shoemaker, Dec. 14, 1961, DP.
7. Darling to Dan W. Turner, Dec. 8, 1961, DP.
8. Darling to Paul H. Appleby, Dec. 14, 1961, DP.
9. Darling to Max McGraw, Dec. 29, 1961, DP.
10. Merle Houts Strasser to Elmer T. Peterson, Dec. 30, 1961, DP.
11. Darling to Mrs. Paul C. Howe, Jan. 30, 1952, property of Mrs. Howe.
12. Edmundson Art Museum folders, DP.
13. David Fyten, Jim Demetrion's Art Empire, Des Moines *Sunday Register,* Nov. 6, 1977, p. 3-B.
14. Darling to George W. Dulany, Jr., June 19, 1941, DP; Darling to Hendrik Willem van Loon, May 26, 1942, DP.
15. Darling to F. T. Schwob, Jan. 30, 1945, DP.
16. Darling to Fred J. Poyneer, Jan. 10, 1942, DP.
17. Darling to A. F. Allen, May 17, 1920, AP.
18. Darling to Mr. and Mrs. C. R. Gutermuth, Aug. 18, 1946, DP.
19. Darling to Leslie Dunlap, July 5, 1961, DP.
20. Dunlap to Darling, June 28, 1961, DP.
21. History of Education in Nine Cartoons Drawn for SUI Library Building, Des Moines *Sunday Register Picture Magazine,* May 31, 1953, pp. 8–9; A Cartoonist Decorates a Building, *Look,* June 16, 1953.
22. Darling to Poyneer, Aug. 20, 1951, DP.
23. Des Moines *Tribune,* Feb. 12, 1962, p. 1; Des Moines *Register,* Feb. 13, 1962, p. 1.

## CHAPTER 22

1. Harlan Miller, Over the Coffee, Des Moines *Register,* Feb. 16, 1962, p. 20.
2. Objectives and Goals of the J. N. "Ding" Darling Foundation, Inc., copy of typescript, 1975; Program, Official Dedication of the J. N. "Ding" Darling National Wildlife Refuge, 1978.
3. Press release from Darling Foundation, Oct. and Nov. 1962.
4. Sherry R. Fisher, In the Footsteps of Lewis and Clark, *Vista,* Fall 1977, pp. 33–36.
5. Objectives and Goals, Darling Foundation.
6. The Etchings of J. N. "Ding" Darling, Iowa State Center, Iowa State Univ., 1975.
7. Ralph Hollander, Ding: Sheldon Exhibit Features Famous Cartoonist, Sheldon and Sibley (Iowa) *Review,* Dec. 3, 1977, p. 3-A; John M. Henry and David L. Lendt, J. N. Ding, photo album of Veishea exhibition of Darling cartoons, Apr. 27–May 26, 1977.
8. Jay N. Darling and John M. Henry, *Ding's Half Century* (New York: Duell, Sloan and Pearce, 1962), p. ix.
9. Ray Kennedy, Eden Fights Back, *Sports Illustrated,* Feb. 3, 1975, pp. 28–30, 33–35; If You're Traveling: J. N. "Ding" Darling National Wildlife Refuge, Sanibel Island, Florida, *Ark,* Spring 1975, p. 16.
10. James Risser, Ways Sought to Halt Loss of Farmland, Des Moines *Sunday Register,* Aug. 7, 1977, p. 1-A; Conservationists End Meeting, Urge "Earth-Care," Des Moines *Register,* Dec. 2, 1976, p. 8-A.
11. Don Muhm, Official Says Soil Erosion Worst Since Depression, Des Moines *Register,* Oct. 8, 1977, p. 4-S.
12. Otto Knauth, Corps Projects on Missouri, "a Disaster," Des Moines *Sunday Register,* Nov. 14, 1976, pp. 1-A, 12-A; Otto Knauth, Clarion Call for Formal Survey of Precious Iowa, Des Moines *Register,* Dec. 14, 1977, p. 6-A.
13. Wildlife Groups Criticize Use of Military Lands, Des Moines *Register,* Nov. 24, 1977, p. 5-A.
14. Otto Knauth, Conservation Officials Lash Critics in Legislature, Des Moines *Register,* Dec. 9, 1977, p. 6-A; Bonnie Wittenburg, Car Official Can Hunt Free, Des Moines *Register,* Oct. 27, 1977, p. 12-A.

15. Larry Stone, Urge Kansas Prairies Be Saved as U.S. Park, Des Moines *Sunday Register,* Sept. 11, 1977, p. 8-D; Otto Knauth, Study Asked for U.S. Park in West Iowa, Des Moines *Sunday Register,* Dec. 25, 1977, p. 1-A; Boundary Waters Protection Near, Iowa PIRG *Gazette,* Fall 1977, p. 1; Compromise on BWCA, Des Moines *Register,* Jan. 2, 1978, p. 10-A; Muhm, Soil Erosion Worst; Land, Water and Energy in Century III, Public Affairs Pamphlet, Iowa State Univ. Ext. Serv., 1977.

16. Darling Duck Stamps Raise Millions for Wetlands Purchase, Sanibel and Captiva (Fla.) *Island Reporter,* Nov. 21, 1975, p. A-24.

17. Scrapbooks, DP; Lawrence B. Fletcher to Darling, Nov. 1, 1944, DP; George Mills, *Harvey Ingham and Gardner Cowles, Sr.: Things Don't Just Happen* (Ames: Iowa State Univ. Press, 1977), p. 121.

18. John M. Henry, The Iowa Award, undated copy of typescript.

19. Foreign Policy Association, *A Cartoon History of United States Foreign Policy: 1776-1976* (New York: Morrow, 1975), pp. 49, 51, 55, 57, 59, 62, 65, 84, 85, 87.

# Selected Bibliography

A. F. Allen Papers, Special Collections Department, University of Iowa Library, Iowa City.
Jay N. Darling Collection, Special Collections Department, Cowles Library, Drake University, Des Moines, Iowa.
Jay N. Darling Papers, Special Collections Department, University of Iowa Library, Iowa City.
This collection includes correspondence up to 1949, scrapbooks, photographs, and other memorabilia.
Jay N. Darling Papers, personal library of Mr. and Mrs. Richard B. Koss, Des Moines, Iowa.
This collection includes correspondence from 1949 forward, family biographies, and miscellaneous conservation publications, some of which are out of print.
Paul Errington Papers, Department of Special Collections, Iowa State University Library, Ames.
General Records, U.S. Fish and Wildlife Service, 1900–1956, National Archives and Records Service, Washington, D.C.
Izaak Walton League of America (Iowa) Papers, Department of Special Collections, Iowa State University Library, Ames.
Christian Petersen Papers, Department of Special Collections, Iowa State University Library, Ames.
President's Committee on Wildlife, January–July 1934, National Archives and Records Service, Washington, D.C.
Everett B. Speaker Papers, Department of Special Collections, Iowa State University, Ames.

INTERVIEWS

Barnes, Walter C. Des Moines, Iowa, Dec. 3, 1977.
Bressman, Earl N. Moorhead, Iowa, Apr. 3, 1977. (Interview conducted by Mrs. Lorna Wessel.)
Clark, Vernon L. Des Moines, Iowa, Jan. 12 and June 25, 1977.
Cowles, Gardner (Mike), Jr. Des Moines, Iowa, Nov. 30, 1977.

Fisher, Sherry R. Des Moines, Iowa, Nov. 16, 1977.
Fox, Rodney T. Ames, Iowa, Oct. 27, 1977.
Gutermuth, Clinton R. (Interview conducted by Elwood R. Maunder and published in *Clinton R. Gutermuth: Pioneer Conservationist and the Natural Resources Council of America*. Santa Cruz, Calif.: Forest History Society, 1974.)
Henry, John M. Des Moines, Iowa, Dec. 22, 1976, and Feb. 24 and Nov. 16, 1977.
Hunter, Mrs. Carolyn. Des Moines, Iowa, Nov. 16 and Dec. 1, 1977.
Huttenlocher, Mrs. Fae. Des Moines, Iowa, Nov. 16, 1977.
Koss, Mrs. Mary Darling. Des Moines, Iowa, Oct. 21, Nov. 4, Nov. 30, Dec. 16, 1976, and Aug. 13 and Dec. 17, 1977.
MacDonald, Kenneth. Des Moines, Iowa, Sept. 7, 1977.
Meaney, Gordon E. Des Moines, Iowa, Sept. 8, 1977.
Parker, Mrs. Addison. Des Moines, Iowa, Sept. 9, 1977.
Strasser, Mrs. Merle Houts. Des Moines, Iowa, June 2, 1977.
Summerfelt, Robert C. Ames, Iowa, Sept. 2, 1977.
Zwart, Mrs. Elizabeth Clarkson. Des Moines, Iowa, Sept. 1 and Oct. 6, 1977.

## LETTERS TO AUTHOR

Colflesh, Mrs. Chloris. Jan. 13 and Jan. 27, 1978.
Cowles, Gardner (Mike), Jr. June 27, 1977.
Goode, D. J. Feb. 24, 1978.
Graham, Mrs. Elizabeth. Oct. 8, 1977.
Henry, John M. Jan. 1, Jan. 6, Jan. 7, Jan. 12, Feb. 17, Mar. 18, Apr. 11, Apr. 30, May 25, June 3, June 16, June 21, July 8, July 11, July 13, July 15, Aug. 12, Aug. 19, Aug. 21, Sept. 17, Oct. 13, Oct. 15, Oct. 26, and Dec. 11, 1977.
Howe, Mrs. Paul C. Sept. 10 and Sept. 20, 1977.
Huttenlocher, Mrs. Fae. Nov. 11, 1977.
Irrmann, Robert H. Jan. 5 and Oct. 24, 1977.
Janss, Mrs. Peter W. Sept. 2, 1977.
Lammers, Ms. Alice. Mar. 8, 1978.
Meaney, Gordon E. Aug. 30 and Sept. 17, 1977.
Paluka, Frank. Oct. 5, 1977.
Parnham, Harold J. Sept. 3, 1977.
Reid, Whitelaw. Mar. 23, and July 11, 1978.
Rinker, Mrs. Oliver Doran. Oct. 14, 1977.
Simpson, Ms. Roberta. Oct. 27, 1977.

## SPEECHES

Darling, J. N. Conservation Education: Observations by Jay N. Darling. Typescript of address dated Sept. 15, 1950.
————. Conservation and Planning. Abstract of address delivered at Lakes Region Planning Institute, Emmetsburg, Iowa, July 10, 1936.
————. Darling Talk at N.Y. Dinner. Undated typescript marked "Confidential, Not for Publication."
————. The Future of Migratory Waterfowl. Speech delivered at the Thirteenth Annual Convention of the Izaak Walton League of America, Chicago, Ill., Apr. 12, 1935.
————. Lee County Mosquito Control. Undated carbon copy of typescript.

_____. Message from J. N. (Ding) Darling to the Annual Banquet, National Wildlife Federation, Miami, Fla., Mar. 19, 1952. Copy of typescript.

_____. Untitled address. Jay N. Darling at Rochester, Minn. Aug. 1, 1946. Typescript of address bears typed note on covering page reading "Manfert Johnson—Kiwanis Club."

_____. What Can I Do for Wild Life Conservation? Undated carbon copy of typescript.

Fisher, Sherry R. J. N. "Ding" Darling and North Dakota. Undated photocopy of typescript.

Henry, John M. Editing Ding's Cartoons. Speech to Des Moines chapter of Sigma Delta Chi. Undated copy of typescript.

MacDonald, Kenneth. Jay N. Darling Dinner. Copy of typescript of address delivered Aug. 22, 1960.


NEWSPAPERS

Des Moines *Capital,* 1911–1913. Newspaper Collection, Iowa State Department of History and Archives, Des Moines.

Des Moines *Register,* 1916–1949. Newspaper Collection, Iowa State Department of History and Archives, Des Moines.

Des Moines *Register and Leader,* 1906–1916. Newspaper Collection, Iowa State Department of History and Archives, Des Moines.

New York *Globe and Commercial Advertiser,* 1911–1913. Microfilm, Library of Congress, Washington, D.C.

New York *Herald Tribune,* 1916–1949. Microfilm, Iowa State University Library, Ames.

Sioux City *Journal,* 1900–1906. Newspaper Collection, Iowa State Department of History and Archives, Des Moines.


BOOKS

Byrnes, Gene, and Bishop, A. Thornton, eds. *Commercial Art: A Complete Guide to Drawing, Illustration, Cartooning and Painting.* New York: Simon and Schuster, 1952.

Carhart, Arthur H. *Water—Or Your Life,* rev. ed. Philadelphia and New York: Lippincott, 1959.

Darling, J. N. *Aces and Kings: Cartoons from the Nebraska State Journal.* Lincoln: Nebraska *State Journal,* 1918.

_____. *Cartoons from the Pen of Jay N. Darling "Ding".* Des Moines: Register and Leader Co., undated.

_____. *Condensed Ink, an Iowa Breakfast Food: Being Cartoons from The Register and Leader, of Des Moines by "Ding".* Des Moines: Register and Leader Co. [1910].

_____. *The Cruise of the Bouncing Betsy: A Trailer Travelogue.* New York: Stokes, 1937.

_____. *Ding Goes to Russia.* New York and London: McGraw-Hill, 1932.

_____. *In Peace and War: Cartoons from the Des Moines Register.* Des Moines: Register and Tribune Co., 1916.

_____. *It Seems like Only Yesterday.* Des Moines: Ding Darling [1960].

_____. *Our Own Outlines of History for 1921 & 1922: Cartoons by J. N. Darling,* Book 8. Des Moines: Des Moines Register and Tribune Co., 1922.

Darling, J. N., and Henry, John M. *As Ding Saw Hoover*. Ames: Iowa State College Press, 1954.

_____. *Ding's Half Century*. New York: Duell, Sloan and Pearce, 1962.

Darling, Marcellus Warner. *Fusing Truth into Life*. Glencoe, Ill.,: Marcellus Warner Darling, 1906.

_____. *Marcellus Warner Darling: 1844–1913*. Glencoe, Ill.: Clara W. Darling [1913].

Ding, J. N. *Cartoons from the Files of The Register and Leader*, 2nd ed. Des Moines: Register and Leader Co., 1909.

_____. *The Education of Alonzo Applegate and Other Cartoons by J. N. Ding*, 2nd ed. Des Moines: Register and Leader Co., 1910.

Foreign Policy Association. *A Cartoon History of United States Foreign Policy: 1776–1976*. New York: Morrow, 1975.

Garraty, John A. *The American Nation: A History of the United States*, 3rd ed. New York: Harper and Row, 1975.

Laycock, George. *The Sign of the Flying Goose: The Story of Our National Wildlife Refuges*, rev. ed. Garden City, N.Y.: Anchor Press/Doubleday, 1973.

Mills, George. *Harvey Ingham and Gardner Cowles, Sr.: Things Don't Just Happen*. Ames, Iowa State Univ. Press, 1977.

Murphy, Dan, ed. *Midwest Farming as Portrayed by a Selection from Ding's Cartoons*. Des Moines: Pioneer Hi-Bred Corn Co., 1960.

Roosevelt, Elliott, ed. *F.D.R.: His Personal Letters, 1928–1945*, vol. 3. New York: Duell, Sloan and Pearce, 1950.

Sears, Paul B. *Deserts on the March*. Norman: Univ. Oklahoma Press, 1935.

Trefethen, James B. *Crusade for Wildlife: Highlights in Conservation Progress*. Harrisburg, Pa.: Stackpole; New York: Boone and Crockett Club, 1961.

U.S. Department of Agriculture. *Report of the Chief of the Bureau of Biological Survey, 1934*. Washington, D.C.: USGPO, 1934.

Vogt, William. *Road to Survival*. New York: Wilff, 1948.

Wallace, Henry A. *The Century of the Common Man*, Russell Lord, ed. New York: Reynal and Hitchcock, 1943.

White, William Allen. *The Autobiography of William Allen White*. New York: Macmillan, 1946.

Willkie, Wendell. *This Is Wendell Willkie*. New York: Dodd, Mead, 1940.

## JOURNAL ARTICLES

Briley, Ronald F. Smith W. Brookhart and Russia. *Annals of Iowa* 42(Winter 1975):541–56.

Brown, Robert Z. United States Water Supply vs. Population Growth. *Population Bulletin* 17(Aug. 1961):85–108.

Darling, Jay N. Conservation: A Typographical Error. *Review of Reviews* 44(Nov. 1936):35–37.

Gutermuth, C. R., and Maunder, Elwood R. Origins of the Natural Resources Council of America: A Personal View. *Forest History* 17(Jan. 1974):4–17.

Haugen, Arnold O. History of the Iowa Cooperative Wildlife Research Unit. *Iowa Academy of Science* 73(1966):136–45.

Henry, John M. A Treasury of Ding. *Palimpsest* 53(Mar. 1972):81–178.

Mills, George. The Des Moines Register. *Palimpsest* 30(Sept. 1949):273–304.

Purcell, L. Edward. Commentary: Will the Real Ding Please Stand Up? *Palimpsest* 57(Jan.–Feb. 1976):30–31.

Rader, Frank J. Harry L. Hopkins, The Ambitious Crusader: An Historical Analysis of the Major Influences on His Career. *Annals of Iowa* 44(Fall 1977):83–102.

Reuss, Carol. Edwin T. Meredith: Founder of *Better Homes and Gardens. Annals of Iowa* 42(Spring 1975):609–19.

## NEWSPAPER ARTICLES

Cartoonist's Works Displayed. *Iowa State Daily,* Apr. 27, 1977, p. 10.

Conservationists End Meeting, Urge "Earth-Care." Des Moines *Register,* Dec. 2, 1976, p. 8-A.

Darling Collection Opened. *Iowa Stater,* Mar. 1976, p. 2.

Darling Duck Stamps Raise Millions for Wetlands Purchase. Sanibel and Captiva (Fla.) *Island Reporter,* Nov. 21, 1975, p. A-24.

Darling Etchings Adorn Gallery. *Iowa Stater,* Jan. 1976, p. 4.

Darling's Cartoon Story of Education Placed. *Daily Iowan,* June 2, 1953.

Ding Honored at Dinner. Des Moines *Register,* Aug. 23, 1960, p. 3.

Dougherty, Steve. A World-famous Cartoonist and Avid Conservationist Who Chronicled Turbulent Events in History Was . . . Ding Darling. Fort Myers *News-Press,* Nov. 21, 1976, pp. 1-D, 5-D.

Flansburg, James. Kenneth MacDonald Retires from R and T. Des Moines *Sunday Register,* Jan. 2, 1977, pp. 1-A, 18-A.

Fyten, David. Jim Demetrion's Art Empire. Des Moines *Sunday Register,* Nov. 6, 1977, p. 3-B.

The Gators are Gasping as Okefenokee Dries Up. Des Moines *Register,* July 19, 1977, p. 8-B.

Gift of Valuable Wildlife Etchings—Another "Ding" Darling Link. *Iowa Stater,* Apr. 1976, p. 11.

Knauth, Otto. Corps Projects on Missouri "a Disaster." Des Moines *Sunday Register,* Nov. 14, 1976, p. 1-A.

MacQuarrie, Gordon. Right Off the Reel. Milwaukee *Journal,* Mar. 7, 1937.

Margaret White Recalls 31 Years at 'Tween Waters. Sanibel and Captiva (Fla.) *Island Reporter,* July 15, 1977, p. B-13.

Muhm, Don. Official Says Soil Erosion Worst Since Depression. Des Moines *Register,* Oct. 8, 1977, p. 4-S.

Risser, James. Ways Sought to Halt Loss of Farmland. Des Moines *Sunday Register,* Aug. 7, 1977, p. 1-A.

Saving the Upper Iowa. Des Moines *Register,* Nov. 22, 1976, p. 8-A.

Scientists Guard Wildlife of Iowa. *News of Iowa State,* Jan. 1960, p. 5.

Sutton, Horace. Captiva Island: Florida's Version of Tahiti Comes Out of Its Shell. Los Angeles *Times,* Oct. 3, 1976, p. 1.

Tuttle, Ries. Darling and Disney Honored. Des Moines *Tribune,* Oct. 12, 1961.

Twombly, Mark. J. N. Ding. Sanibel and Captiva (Fla.) *Island Reporter* Apr. 29, 1977, pp. 1-B, 14-B, 15-B.

Yergin, Daniel. A Talented Man—Henry Wallace Has Remained a Shadowy Figure. Des Moines *Sunday Register,* Dec. 19, 1976, p. 4-B.

## MAGAZINE ARTICLES

A Cartoonist Decorates a Building. *Look,* June 16, 1953.

Bush, Monroe. Silent Spring . . . Noisy Autumn. *American Forests,* Oct. 1962, pp. 12, 52–54.

'Bye Now—It's Been Wonderful. *Life*, Feb. 23, 1962, p. 44.

Clepper, Henry. What Conservationists Think about Conservation. *American Forests*, Aug. 1975, pp. 28–29, 46–47.

Conrad, Paul. F. The Ticklish Art. *Iowa Alumni Review*, Dec. 1963, p. 5.

Cottam, Clarence. In Memory of Ding. *Michigan Conservation*, May–June 1962, pp. 33–35.

Darling, J. N. A Duck for Every Puddle. *Look*, Aug. 19, 1947.

———. Last Hope for Ducks? *Outdoor America*, Dec. 1960, pp. 2–3.

———. Speaking of Flood Control. *Public Service Magazine*, Apr. 1945, pp. 21–25.

———. The Story of the Wildlife Refuge Program, Part I. *National Parks Magazine*, Jan.–Mar. 1954, pp. 6–10, 43–46.

———. Underground Water More Precious Than Gold. *Wyoming Wildlife*, Mar. 1962, pp. 11–15.

———, as told to Boyden Sparkes. Washington Wild Life. *Saturday Evening Post*, Sept. 21, 1935, pp. 14–15, 61, 62, 64, 65.

"Ding" Darling's 42 Years of Cartooning. *Quill*, Jan. 1943.

"Ding's" Genial Pen Refused to Libel Life. *Editor and Publisher*, Aug. 9, 1924.

Farb, Peter. Hugh Bennett: Messiah of the Soil. *American Forests*, Jan. 1960, pp. 18–19, 40, 42.

Gabrielson, Ira N. Ducks Can't Lay Eggs on Picket Fences. *American Forests*, Sept. 1962, pp. 16–21, 54–55.

History of Education in Nine Cartoons Drawn for S.U.I. Library Building. Des Moines *Sunday Register Picture Magazine*, May 31, 1953, pp. 8–9.

Hume, A. H. The Playhouse of a Cartoonist. *House Beautiful*, Apr. 1924, pp. 482–84.

If You're Traveling: J. N. "Ding" Darling National Wildlife Refuge, Sanibel Island, Florida. *Ark*, Spring 1975, p. 16.

Jay N. "Ding" Darling (1876–1962). *Ducks Unlimited*, Jan.–Feb., 1975, back cover.

Kennedy, Ray. Eden Fights Back. *Sports Illustrated*, Feb. 3, 1975, pp. 28–30, 33–35.

Lost Darling. *Newsweek*, Apr. 25, 1949, p. 60.

Mahoney, Tom. How to Be a Cartoonist. *Saturday Evening Post*, Oct. 19, 1940, p. 34.

More Ding Cartoons. *American Forests*, Oct. 1962, pp. 16–21, 54–55.

Original "Darling" Artwork Decorates Headquarters. *Gardener*, Nov.–Dec., 1977, p. 5.

Paluka, Frank. The S.U.I. Ding Darling Cartoon Collection. *Iowa Alumni Review*, Dec. 1962, pp. 7–9.

Proper, Carl C. "Ding," The World's Greatest Cartoonist. *People's Popular Monthly*, Oct. 1923, pp. 7–39.

So Long, Ding. *American Forests*, Mar. 1962, p. 39.

Soucie, Gary. Will Sanibel Dreams Thwart Get-Rich Schemes? *Audubon*, Mar. 1976, pp. 114–16.

Trefethen, James B. The Wildlife Management Institute. *New York Conservationist*, Dec.–Jan. 1960–1961, pp. 14–15.

Wildlife Conference. *Time*, Feb. 27, 1939, p. 41.

Yes, Virginia, There Is a "Ding." *Iowan*, Oct.–Nov., 1959, pp. 38–48.

## PAMPHLETS

Darling, J. N. Poverty or Conservation? National Wildlife Federation, 1944.

———. The Story of the Underground Water Table. *Audubon Nature Bulletin* (formerly *School Nature League Bulletin*), Ser. 15, No. 1, 1962.

———. Why, How and What. Undated prospectus outlining need for Conservation Clearing House.

"Ding" on Cartooning. J. N. "Ding" Darling Foundation, 1961.

Duck Stamps and Wildlife Refuges. U.S. Dept. Int., Fish and Wildlife Serv., Circ. 37, 1956.

Edge, Rosalie. The United States Bureau of Destruction and Extermination: The Misnamed and Perverted "Biological Survey." Emergency Conservation Committee, Sept. 1934.

The Etchings of J. N. "Ding" Darling. Iowa State Center, Iowa State Univ. [1975].

50th Anniversary: Technical Journalism at Iowa State College, 1905–1955. Dept. Tech. J., Iowa State College, 1955.

Lower Souris National Wildlife Refuge. U.S. Dept. Int., Fish and Wildlife Serv., Mar. 1960.

Malheur National Wildlife Refuge. U.S. Dept. Int., Fish and Wildlife Serv., Nov. 1959.

The Nature Conservancy News. Vol. 10, No. 4, Oct. 1960.

Okefenokee National Wildlife Refuge. U.S. Dept. Int., Fish and Wildlife Serv., Refuge Leaflet 8, 1954.

Sand Lake National Wildlife Refuge. U.S.Dept. Int., Fish and Wildlife Serv., Refuge Leaflet 32, undated.

Shoemaker, Carl D. The Stories behind the Organization of the National Wildlife Federation and Its Early Struggles for Survival. National Wildlife Federation, 1960.

Suggested Plans and Information for Use in the Formation of County and State Wildlife Federations. General Wildlife Federation [1936].

Upper Souris National Wildlife Refuge. U.S. Dept. Int., Fish and Wildlife Serv., Refuge Leaflet 35, undated.

## DIRECTORIES

*Annals of Iowa,* Vol. 30, Jan. 1950.

*Current Biography,* Vol. 3, No. 7, July 1942.

*Jay N. Darling Papers: Index.* Univ. Iowa Library, Dept. Special Collections.

*The Story of Iowa,* Vol. 3. New York: Lewis Historical, 1952.

*Who's Who in Iowa.* Des Moines: Iowa Press Assoc., 1940.

Yates, Stanley. *A Guide to the Manuscript Collections in the Iowa State University Library,* Apr. 1977.

## MISCELLANEOUS

Darling, Jay N. Objectives and Functions of the American Wildlife Institute. Undated copy of typescript.

Gordon, Seth. Ding Darling: Dynamic Conservation Leader. Undated and unpublished copy of typescript of article originally written for National Rifle Association.

Henry, John M. Ding Things. Copy of typescript, Apr. 4, 1977.

Huttenlocher, Mrs. Fae. Brief History of the MGC of Des Moines (from notes furnished by Mrs. Forest Huttenlocher). Undated typescript. Men's Garden Clubs of America, Des Moines, Iowa.

An Invitation from the President of the United States. Prospectus for North American Wildlife Conference, Feb. 3–7, 1936 [1935].

Minutes of Darling Dinner Meeting Held at the Waldorf-Astoria, New York City, Apr. 24, 1935. Original typescript.

# Index

199